VICTORIAN JESUS: J.R. SEELEY, RELIGION, AND THE CULTURAL SIGNIFICANCE OF ANONYMITY

ECCE HOMO

A SURVEY OF

THE LIFE AND WORK OF JESUS CHRIST

'Auctor nominis ejus Christus Tiberio imperitante per procuratorem Pontium Pilatum supplicio affectus erat' TACIT. *Ann.* l. 15

FIFTH EDITION, WITH NEW PREFACE

London

MACMILLAN AND CO.

1866

(Right of translation and reproduction reserved)

The title page for *Ecce Homo: A Survey in the Life and Work of Jesus Christ*, 5th edn (London: Macmillan, 1866). Courtesy of the University of Queensland.

IAN HESKETH

Victorian Jesus: J.R. Seeley, Religion, and the Cultural Significance of Anonymity

UNIVERSITY OF TORONTO PRESS
Toronto Buffalo London

Toronto Buffalo London
www.utppublishing.com

ISBN 978-1-4426-4577-6

Studies in Book and Print Culture

Library and Archives Canada Cataloguing in Publication

Hesketh, Ian, 1975–, author
Victorian Jesus : J.R. Seeley, religion, and the cultural significance of anonymity / Ian Hesketh.

(Studies in book and print culture)
Includes bibliographical references and index.
ISBN 978-1-4426-4577-6 (hardcover)

1. Seeley, John Robert, Sir, 1834–1895. 2. Seeley, John Robert, Sir, 1834–1895. Ecce Homo. 3. Historians – Great Britain – Biography. 4. Anonymous writings, English – History and criticism. 5. Authors and publishers – Great Britain – History – 19th century. 6. Publishers and publishing – Great Britain – History – 19th century. 7. Religion and literature – Great Britain – History – 19th century. 8. Literature and society – Great Britain – History – 19th century. 9. Jesus Christ – In literature. I. Title. II. Series: Studies in book and print culture

DA3.S4H47 2017 907.2'02 C2017-902182-6

This book has been published with the help of a grant from the Federation for the Humanities and Social Sciences, through the Awards to Scholarly Publications Program, using funds provided by the Social Sciences and Humanities Research Council of Canada.

University of Toronto Press acknowledges the financial assistance to its publishing program of the Canada Council for the Arts and the Ontario Arts Council, an agency of the Government of Ontario.

Canada Council for the Arts Conseil des Arts du Canada

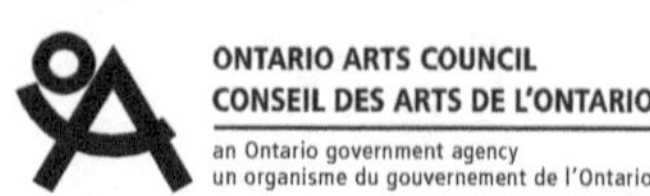

Funded by the Government of Canada Financé par le gouvernement du Canada Canada

For my wonderful parents,
Chris Turner and Bob Hesketh

Contents

List of Illustrations

Acknowledgments

The possibility of this book first occurred to me on a research trip to the UK in 2007 when visiting John Seeley's archive at Senate House Library, University of London. I was reading through Seeley's correspondences for a book that became *The Science of History in Victorian Britain* (2011) when I came across a folder with material relevant to the publishing of Seeley's anonymous *Ecce Homo*. I recognized immediately that these documents shed light on an incredible story that had yet to be told. I originally envisioned a slim book that would narrate the controversy surrounding the publishing of *Ecce Homo*, but further research led me to realize that this microhistory could be expanded to explore larger themes in Victorian religious history as well as in the history of anonymous publishing. It has now been ten years since my original discovery, and the book that you hold in your hands is the result.

The early research was supported by a Social Sciences and Humanities Research Council of Canada Postdoctoral Fellowship that I held at the University of British Columbia from 2006 until 2008. I also received an Early Career Research Grant from the University of Queensland in 2012 that allowed me to follow up on my earlier trip while widening my research base. I must also acknowledge the Publication Grant provided by the Awards to Scholarly Publications Program in the Federation for the Humanities and Social Sciences.

I am particularly grateful to the many archivists and librarians who have assisted me throughout the ten years that I have been at work on this project, especially at the Senate House Library, University of London, and the British Library, where I spent much time in the manuscript and newspaper reading rooms. I also had the help of a research assistant, Shelise Robertson, who worked with me in transcribing the Seeley–Sidgwick correspondence.

I've been very lucky over the years to learn a great deal from a wonderful group of colleagues who were willing to discuss and read parts of this book when it was very much a work in progress. At the University of British Columbia, Okanagan campus, James Hull and Oliver Lovesey were enthusiastic supporters of the project when it was in a very early stage. At Queen's University, Kingston, Daniel Woolf was always willing to impart his vast knowledge of the history of historical writing, which I greatly appreciated. And at the Institute for Advanced Studies in the Humanities (formerly the Centre for the History of European Discourses) at the University of Queensland, my work on this project received extensive feedback from many of my current and former colleagues, including Peter Harrison, Leigh Penman, Ian Hunter, Knox Peden, Nicholas Heron, Karin Sellberg, Michael Ostling, Simon During, James Ungureanu, Peter Cryle, Marina Bollinger, Kim Hayek, Pete Jordan, Phil Almond, Daniel Midena, and Gary Ianziti. I am particularly thankful to Leigh Penman, who spent countless hours discussing all things Seeley with me, while helping me with some of the methodological issues that are at the heart of the book. And Simon During read through the complete first draft of the book manuscript, providing a list of important criticisms that formed the basis of my early revisions. Others who have provided helpful critiques and discussions at conferences or via email include Bernard Lightman, Anne DeWitt, Efram Sera-Shriar, Conal Condran, Richard Yeo, Todd Webb, Geoff Read, Robert Griffin, Patrick Low, Andreas Sommer, Rosemary Mitchell, Helen Kingstone, Mary Isbell, Geoff Belknap, and Robert Patten.

I cannot let this opportunity pass without making special acknowledgment of Leslie Howsam. She was very encouraging about the project from the beginning and helped secure the interest of the University of Toronto Press. She was moreover always willing to respond to my many, many emails querying a variety of subjects related (and sometimes not!) to the book. And, of course, her work on publishers and Victorian historical writing provided the initial inspiration for the book as well as the methodological approach that has been taken.

It has also been a pleasure working again with the University of Toronto Press. My first editor, Siobhan McMenemy, who believed in the project at a very early stage, worked quickly once the manuscript was finally submitted to secure timely and insightful reader reports, and perhaps more importantly helped me to see the larger significance of the project that was made apparent by those reports. Mark Thompson took over editing duties and was very prompt in shepherding the manuscript through the latter stages of the evaluation process. He did an excellent job getting all the acceptances and approvals that are necessary when publishing with a prestigious university press in Canada. The book was copyedited by Judith Williams, whose incredible attention to detail saved me from making several embarrassing errors. And Ruth Pincoe did a fabulous job with the index.

I'm also extremely grateful to the two anonymous referees of the manuscript. They gave such wonderful critical analyses that it is impossible to summarize just how the manuscript was improved by their careful readings. They brought to light both problems and possibilities that made very clear the particular direction the manuscript needed to take as it was revised into its current published state. All remaining errors are, however, entirely of my own making.

Finally, I'd like to thank my family. The book is dedicated to my parents, particularly for their continuing support of my chosen profession, which seems only to take me farther and farther away from home. Other than myself, my partner Cleo has had to live with this project more than anyone, and I have always appreciated her willingness to at least feign interest at my latest discovery concerning the story of *Ecce Homo*. She even went through some of the manuscript and gave me excellent suggestions for making it more readable. Our daughter Sadie has also been a wonderful inspiration. Her excitement upon learning new things is infectious.

VICTORIAN JESUS: J.R. SEELEY, RELIGION, AND THE CULTURAL SIGNIFICANCE OF ANONYMITY

Prologue

The Forgotten Story of *Ecce Homo*

Every act of reading is an act of forgetting …

– James Secord, *Victorian Sensation* (2000)

As legend has it, sometime in the late winter of 1866 the gregarious publisher Alexander Macmillan invited sixteen of London society's leading intellectual and literary figures to his home to meet the unknown author of the anonymous *Ecce Homo: A Survey in the Life and Work of Jesus Christ* (1865) (see frontispiece). This would have been a much-anticipated event, as *Ecce Homo* was fast becoming the sensation of the literary season. Unfortunately for Macmillan's guests, however, the author, while present, was never revealed as such and kept silent about his authorship. As the story goes, everyone left the dinner party none the wiser about who among them was the author.[1]

About the time that this notorious dinner party took place, the poet and future author of *Culture and Anarchy* (1869), Matthew Arnold, wrote to his mother that "[e]very one is beginning to talk of a new religious book, called Ecce Homo."[2] It was, moreover, a book that was often difficult to purchase, as the bookshops sold out of their copies immediately on every new printing. To meet the ever-increasing demand, Macmillan resorted to some unorthodox methods. The London correspondent for the *Bury and Norwich Post* cited, as "proof of the popularity of the fascinating, but extremely dangerous … *Ecce Homo*," the fact that it was being sold at train stations. While waiting for a train at King's Cross, he watched "Smith and Son's boys walking up and down the platform the length of the train, shouting '*Daily Telegraph, Standard, Morning Star,* ECCE HOMO! … *Ecce Homo,* second edition!'" Upon being asked about the book by a potential customer, one of the boys replied, "a capital book, sir; selling a great many, sir."[3] The correspondent found this to be astonishing.

The buzz surrounding the book was reflected in the periodical press, where reviewers heatedly debated the way in which Jesus was presented, not as a worker of miracles but rather as the founder of a society based on a universal moral code. It was, according to *Ecce Homo*, Christ's "enthusiasm for humanity" that led to the creation of a Christian society that integrated religious, moral, and political life. In this way *Ecce Homo* avoided divisive doctrinal and theological issues, putting them to one side for a future study, in order to focus on Christ the man rather than Christ the son of God. While reviewers debated whether such a division in Christ's life, even if it was simply used as a heuristic, was wise or even possible, another debate emerged in the bookshops and in private conversations about the book's author. Who was he? Indeed, as exciting and disturbing as the content of the book was the mystery surrounding its authorship.

If the content of the book was what initially got people talking, the intense speculation about its authorship helped make the book a full-blown literary sensation. And that speculation was rampant, in large part because the methodology of the book necessarily veiled the author's own theological views. Readers searched the text to uncover its hidden meanings in order not only to discern the true theology that led to the study in the first place but also to discover some indication of the author's identity. It was variously surmised that the author must be the Catholic John Henry Newman or the politician William Ewart Gladstone or the theologian Arthur Penrhyn Stanley or the historian James Anthony Froude or the novelist George Eliot or perhaps even Emperor Napoleon III. The author was also considered to be more generically a High Churchman, a Low Churchman, a Broad Churchman, a Unitarian, a Catholic, a layman, or an atheist. For some, it was the work of an orthodox genius seeking to unite the various Christian denominations; for others it was the work of the devil. It was even thought to have been "vomited," as one famous evangelical Lord put it, "from the jaws of hell."

Such speculations about the author's identity and true purpose, of course, often said more about the fears and desires of the book's many readers than they did about its reputed author. But encouraging readers to speculate about the author's identity was a marketing strategy that Macmillan cultivated to great success throughout the first year of *Ecce Homo*'s publication. Inviting guests to dine with the great unknown author of *Ecce Homo*, without actually revealing to the guests who the author was among them, was only one of the ways that Macmillan generated interest in the author's identity.

This present study is about the publishing and reception of *Ecce Homo* and focuses particularly on the issue of authorship, which was central to the way the book was conceived, marketed, received, and debated by its many readers and reviewers. The book's almost immediate success was a great surprise to both its publisher and author. Every print run of fifteen hundred to three thousand copies was sold out within just a few weeks of being printed, necessitating new reissues or new "editions," in the language publishers used at the time. *Ecce Homo* went through seven such editions (reissues) in as many months and sold upwards of twenty thousand copies in just two years.[4] In hindsight these seem like astonishing figures for a book with such an off-putting Latin title. But this was a period when "cheap Bibles" were readily available for an ever-expanding reading public that would have been intimately familiar with Pontius Pilate's phrasing in John 19:5, spoken to a hostile crowd before Jesus's crucifixion.[5] *Ecce Homo* or "behold the man" was certainly a suitable title for a book focusing on Christ's character and deeds, but most present-day readers will no doubt be more familiar with Nietzsche's book of the same title, or even more recently of Elías García Martínez's fresco, *Ecce Homo* (c. 1930), that was infamously "restored" by Cecilia Giménez to look, as one scribe has put it, like a "crayon sketch of a very hairy monkey in an ill-fitting tunic."[6] Coming a distant third, then, is the anonymous *Ecce Homo* of 1865, a book that has well and truly been forgotten.

What is astonishing is just how quickly this process of forgetting took place. During 1866, *Ecce Homo* was the most hotly debated book of the full calendar year. It was reviewed in all of the leading quarterly, monthly, and weekly periodicals. And the review literature itself spawned articles summarizing the key debates, while several pamphlets were distributed and several books were written in direct response to *Ecce Homo*.[7] But while this extensive literature debated the book's merits and flaws, much of that debate became entangled in assumptions about the author's masked identity. And when that identity became widely known in November 1866, interest in the book, along with the debate about its meaning, evaporated into the ether. Its author, while certainly not approaching the celebrity status of some of those who were originally thought to have written it, was a known quantity. A book that previously could not be categorized was now, perhaps too easily, aligned with one of the main parties of the Anglican Church, confirming the speculation of many, but frustrating those who hoped that it represented something more, something different, something new.

Ecce Homo would not be entirely forgotten in the nineteenth century, however. The long-promised sequel was eventually published as *Natural Religion* in 1882, a full seventeen years after the original. It was actually serialized before that in *Macmillan's Magazine*, but as the articles were not signed, and moreover did not mention *Ecce Homo*, readers could not have been expected to know that the articles were in any way related. When the book version was published, it was signed "by the author of *Ecce Homo*," thereby indicating its bibliographical lineage and ensuring that it would at least receive some notice. Most reviewers, however, were only interested in what *Natural Religion* indicated about the development of the author's religious views since 1865. Many were, moreover, confused as to why the author would continue to hide behind what was now a fairly transparent veil. And yet the author's name would not be officially assigned to either *Ecce Homo* or *Natural Religion* until posthumous editions were published in 1895. It was only upon the author's death that commentators and obituary writers were able openly to discuss the identity of the author of *Ecce Homo*, who was by this time more widely known as the much-respected Regius Professor of Modern History at Cambridge, Sir John Robert Seeley.

This episode is worth returning to for a variety of interrelated reasons, not least because *Ecce Homo* was such a sudden and controversial literary sensation that captured the public's imagination in profoundly contradictory ways. Moreover, the fact that it was so easily forgotten is an interesting aspect of this story as well, telling us much about the protean and ephemeral nature of religious debate and opinion in this heady period. Indeed, between the publishing of *Ecce Homo* and *Natural Religion*, the conception of Christ presented in the former had become a widely accepted interpretation. But as *Ecce Homo*'s Jesus became the Victorian Jesus, the source for this perspective was largely forgotten as debates about Christianity moved in a much different direction. The fact that *Natural Religion* had very little to say about Christ, and focused instead on the issue of harmonizing science and religion, meant that it had the effect of reinforcing the view that its predecessor was no longer at the forefront of contemporary concerns. Taken together, however, the reception of *Ecce Homo* and its sequel gives us much insight into just how quickly religious views were changing within the context of a rapidly modernizing society.

The publishing of *Ecce Homo* was also a transformative event in the life of an important Victorian historian and public figure, the interest in whom has been growing in recent years. This is not surprising, as Seeley, soon after the controversy of *Ecce Homo*, was appointed to the chair of modern history at Cambridge, and thereafter helped to shape the burgeoning historical profession in England. Not only did he help articulate the dominant mode of historical thinking in the period, arguing that politics and history were intimately connected and that the historian was expected to embrace an empirical and disinterested methodology, unlike the literary men of the discipline's past, he also wrote an influential history of the British empire towards the end of his life.[8] His realistic portrayal of Britain's imperial past, along with his suggestions about how the empire should be shaped in the future, gained many adherents in the upper echelons of England's political culture, leading directly to his knighthood.

As the historian and polemicist of empire, Seeley has understandably gained a great deal of attention from historians of an imperial Britain, from international relations scholars who themselves have recently taken a "historical turn," making Seeley's work doubly relevant, and from historians of historical writing.[9] While *Ecce Homo* has been deemed by many of Seeley's biographers and commentators as relatively important in elucidating a guiding commitment in his thought towards integrating the religious and the political,[10] the publishing and reception of *Ecce Homo* have been almost entirely ignored. And those few historians who have considered its reception were not interested in issues surrounding the nature of its publishing and authorship.

Daniel Pals, for instance, who has situated *Ecce Homo* within a larger context of the developing Victorian genre known as the historical Jesus, views its reception as the result of shifting Church allegiances that were just becoming evident in the mid-1860s. But, apparently, Pals did not see the fact that no name was attached to the book as a significant issue on which to comment.[11] Moreover, in Deborah Wormell's very fine intellectual biography of Seeley, *Ecce Homo* does receive extensive treatment, but there is very little discussion of the intense controversy about its authorship or about how the controversy may have affected Seeley's later, more respectable career.[12]

It is assumed that, because Seeley's authorship became so widely known, it was always widely known. In studies of Seeley's life and work, if *Ecce Homo* is mentioned, it is typically accompanied by the suggestion that Seeley's authorship was an "open secret."[13] But for the space of a full year, when *Ecce Homo* was most explicitly debated, Seeley was not widely believed, never mind confirmed, to be the author.

The authorship was undetermined and the subject of much speculation; moreover, Seeley desperately did not want his secret to be discovered. But because the controversy has always been viewed with the benefit of hindsight – the hindsight of knowing what an important and influential historian of empire Seeley would become – this episode has seemed like an amusing appetizer for those interested in Seeley's apparently more important, mature studies. In order to gain a full understanding of the intimately related issues of authorship and content of this remarkable book, what is needed is an approach that reintroduces to the reader a sense of the contingencies that were at play during the year when the authorship of *Ecce Homo* was unknown. It is only by acknowledging the fact that readers were in the dark about the author's identity that we can make sense of the diverse interpretations that were being debated on a daily basis.

This is not to suggest that Pals and Wormell were wrong to neglect some of these issues; different generations examine the same material with different lines of questioning. When I originally visited Seeley's archive in the Senate House Library at the University of London, I was seeking to answer the questions of an older generation. I wanted to know more about Seeley's historical methodology as it appeared in his *The Life and Times of Stein* (1878) and *The Expansion of England* (1883). What I was not prepared to delve into was the extensive correspondence that has been preserved, particularly the letters from Macmillan, about the publishing and reception of *Ecce Homo*.[14] These letters discuss the early debate surrounding the book and include schemes for orchestrating reviews that would present the book and its author in a more positive light. They also illuminate Macmillan's various marketing tactics for getting readers and friends speculating about the author's identity as well as provide insight into how Seeley's authorship was eventually exposed. These letters led me to the archives of other key figures as well as to several reviews that have yet to be digitized in the growing Victorian periodical databases that would otherwise have been very difficult to find. It has since become clear to me that *Ecce Homo* was an important nexus that connected many friends, colleagues, various schools of religious thought, and readers in a common discussion about Christianity and authorial identity in a way that can tell us much about the controversial religious debates of the mid-Victorian period, as well as the processes involved in publishing a remarkable literary sensation.

I recognized, too, that this material could not be approached by the methods of intellectual history and the history of scientific and religious thought alone, but required the insight and research questions that have been pursued by the interdisciplinary approach of book history, which brings together the methods of history, literature, and bibliography.[15] In this regard, my debt to James Secord's *Victorian Sensation* (2000) will be obvious to many. Secord showed that the *Vestiges of the Natural History of Creation* (1844) – a book that was destined to be overshadowed by Darwin's *On the Origin of Species* (1859) – was absolutely central to the Victorian engagement with evolutionary thought and that its anonymity was integral to its popularity. And he did so not just by describing the ideas contained in *Vestiges* but by focusing on the book's publishing, reception, and authorship. More to the point, Secord brought reading as a category of analysis into the history of science by showing how the many diverse readers of *Vestiges* reimagined the book in their own terms while helping to shape future editions, thereby playing a role in its continual recreation.[16] Moreover, as Bernard Lightman's recent extensive work on the popularization of science indicates, many long-ignored works of popular science from the Victorian period were crucial in shaping public perceptions of science, so much so that scientific luminaries such as Thomas Henry Huxley were often following authorial and publishing trends established by subsequently little known journalists.[17] But given that, as Secord puts it so well, "[e]very act of reading is an act of forgetting,"[18] works that were foundational to particular moments in time all too often get supplanted by those that speak more loudly to the concerns of the present. Book history is an approach that mitigates this presentism while redirecting the historian's focus to moments in time before this process of forgetting took place. Moreover, by tracking the reading of particular works and ideas, book history, when done well, can illuminate just how this process of forgetting occurs. Ultimately, that is what I hope to have achieved with this study of *Ecce Homo*.[19]

Of course, it is not just ideas, books, and authors that get forgotten in the process of reading but publishers as well, who are often ignored as simply playing an enabling role in the production of knowledge. Within book history, however, there is much debate as to whether the first step in the creation of the book should be regarded as a decision made by the publisher rather than the author.[20] We certainly need not go that far in understanding that publishers play a crucial role in the production and reception of books. Leslie Howsam's work, in particular, has brought several Victorian publishers out from behind the curtain to show just how central they were in shaping the form and content of a variety of subjects from natural science to politics to history.[21] Macmillan himself has appeared in several of Howsam's studies as an extremely well connected and capable publisher. She shows that he had a remarkable ability to seek out productive and responsible writers that he could then mould in order to produce works that were both valuable and marketable. That he was able to help shape the pedantic and almost hopelessly tedious Edward Freeman into a viable author of popular history is just one example of Macmillan's remarkable publishing skills at work.[22]

Closely examining the publishing of *Ecce Homo* will therefore provide further insight into Macmillan's processes of publishing, specifically with regard to the surprisingly little examined practice of anonymous publishing. This neglect is surprising because the study of anonymous publishing offers us a privileged way of accessing the role of publishers, since their involvement in the publishing of anonymous works makes their role more visible. Publishers often become the public face for their unknown authors; this was certainly the case with *Ecce Homo*. Even so, much secondary literature on anonymity concerns the personal reasons for masking a given author's identity rather than the important commercial motivations that become apparent when focusing on the publisher.[23] Moreover, very little has been written specifically on anonymous *nonfiction* publishing, as most studies consider the celebrated cases of fictional anonymity and pseudonymity such as Walter Scott, Charles Dickens, and George Eliot.[24]

This study of *Ecce Homo* is therefore intended to fill a rather large gap in our understanding of these issues in regard to the publishing of anonymous theological writing and anonymous nonfiction more generally. While anonymous and pseudonymous book publishing was actually very common in the early nineteenth century,[25] by the middle of the century, it came under increasing scrutiny. This was most clear in the realm of journalism, as the longstanding corporate anonymity of periodicals and newspapers came into conflict with a burgeoning liberal ideology that fetishized the "individual opinion."[26] While the debate over anonymous journalism does not quite map onto the issues surrounding anonymous book publishing, anonymity itself was being viewed with scepticism as at best a marketing ploy and at worst a way for an author to avoid responsibility for his or her ostensibly public views.

Within this context, *Ecce Homo* touched off a debate about the ethics of anonymous book publishing. Lying was believed to be an unfortunate but necessary evil in order to protect an author's anonymous identity, but this rationale seemed particularly problematic with reference to a book that was ostensibly about the foundations of Christian morality. When coupled with the fact that *Ecce Homo*'s authorship became central in marketing the book, the ethical quandaries at the heart of anonymous publishing were exposed in a particularly unflattering light. One commentator summarized the problem quite appropriately, believing that "*Ecce Homo* will become *the* instance in the casuistry of anonymous publishing."[27] And *Ecce Homo* would indeed represent a turning point in the debate about anonymous publishing; it became after 1865 more difficult to defend the practice. And while the practice persisted into the final decades of the nineteenth century,[28] it certainly declined within the realm of the nonfiction book industry. Examining the reception of *Ecce Homo* from this perspective, therefore, helps explain why the practice went into decline.

If anonymous publishing was beginning to be so scrutinized, a pertinent question worth asking is why Seeley chose, seemingly against the tide of liberal opinion, to hide his identity. Previous studies have highlighted the deeply personal issues that were likely involved in Seeley's decision. As will be further discussed in the chapters to follow, Seeley's family was at the centre of evangelical life in London, and Wormell points out that Seeley especially wanted to avoid upsetting his sisters and father.[29] What Wormell failed to consider, however, is that Seeley's father was a well-known evangelical publisher and author, who also happened to write anonymous studies of timely and contentious subjects, though unlike his son he did so in order to promote a literal reading of scripture. While Seeley abandoned evangelicalism in favour of some of the liberal Anglican views that so bristled against his father's faith, being brought up by a publisher and author meant that Seeley was very familiar with both roles as well, and certainly had his own ideas about how *Ecce Homo* should be published and marketed.

As we will see, however, far from initially being a marketing ploy or a way to avoid familial responsibility, *Ecce Homo* was published anonymously, according to Seeley, for the same rationale that underpinned some of the arguments that were being used to make the case for the necessity of a signature. On one hand, he argued that signing his name to the book would necessarily associate it with one of the parties of the Anglican Church, which was something he wanted to avoid. On the other, he argued that his investigation necessitated an entirely disinterested approach, such that he had to shed all of his preconceptions and adopt an identity that was, in a sense, different from the one signified by his name. He wanted the book to be read and judged not on the basis of his identity, but on the basis of his investigation alone. It must be stressed that Seeley was writing about a subject that was not generally open to much interpretation. It therefore followed in Seeley's mind that, for the image of Christ to become open to alternative interpretations, the author had to remain hidden. Otherwise he would be dismissed as simply perpetuating a particular view in conformity with a particular theological line. That a particular view of Christ became quite dominant in light of *Ecce Homo*, that is to say the humanitarian who was most concerned with providing a model for an ethically based politics, goes some way to suggest that Seeley was largely successful in aligning his strategies of authorship with the book's ultimate message, even if it was presented as an ambiguous one.

However, Seeley's strategies of authorship were often at odds with Macmillan's strategies of marketing the book. Indeed, if Seeley seemed to have justifiable methodological and epistemological reasons for writing *Ecce Homo* anonymously, these were quickly lost on a reading public that found itself being encouraged by the publisher to speculate on the author's identity. Given the central role of Macmillan in the publishing and reception of *Ecce Homo*, he is therefore an important character in this story. Macmillan's contribution, which has been discerned largely through his correspondence, provides wonderful insight into the multifaceted and complex publishing world that exists just beyond the printed page. This book is also, therefore, a study of the relationship between publisher and author and how their initially productive working relationship became fractured when the reading public began to seek out and eventually discover *Ecce Homo*'s secrets. While the image that emerges of Macmillan may not be quite as pious or as generally flattering as that presented in older studies,[30] it will become apparent that he was remarkably cunning in his ability to exploit the perceived interests of readers by relying on a diversity of marketing and publishing strategies.

We first turn, however, to the world of religious publishing in the early 1860s in order to contextualize the controversial theological debates that *Ecce Homo* was seeking to transcend upon being published in 1865.

Chapter One

Authority and Authorship

Our duty is to follow the Truth, wherever it leads us, and to leave the consequences in the hands of God.

– J.W. Colenso, *The Pentateuch and Book of Joshua Critically Examined* (1862)

There is truth of science and truth of religion: truth of science does not become truth of religion until it is made to harmonise with it.

– Matthew Arnold, "Dr. Stanley's Lectures on the Jewish Church," *Macmillan's Magazine* (1863)

On 26 January 1866, Trinity College (Cambridge) fellow and future professor of moral philosophy Henry Sidgwick wrote to his mother, "I do not know anything about *Ecce Homo*, except that every one here speaks highly of it." He said that a review of the book did not impress him much, but, given the reception the book was receiving in Cambridge, he realized that he would have to read it eventually.[1] Less than a month later he reported that "*Ecce Homo* is a great work." Even while criticizing some aspects of its method, Sidgwick argued that it "is surprisingly powerful and absorbing, almost sublime in parts. It has made a great sensation here." Indeed, as we will see, it spoke to a particular theological need felt by Sidgwick and many of his peers in Cambridge. Sidgwick found it worth noting one other thing as well: "The author keeps his secret."[2]

There was, of course, a very obvious reason for the author's identity to remain a secret, one that may very well trump the commercial and familial reasons that were also no doubt important and will be explored later in the present study. Simply stated, writing on religious subjects in the 1860s was a highly dangerous activity. That Seeley chose to write on perhaps the most contentious of those subjects, namely Jesus Christ himself, while presenting his analysis as underpinned by an entirely neutral, scientific method, was perilous indeed. As he explained in *Ecce Homo*'s preface, he sought to "trace [Jesus Christ's] biography from point to point, and accept those conclusions about him, not which church doctors or even apostles have sealed with their authority, but which the facts themselves, critically weighed, appear to warrant."[3] These words would have associated the book, in the minds of many readers, with the extension of scientific modes of thought into terrain where other modes of investigation were required.

Indeed, *Ecce Homo* was published during a period of intense religious debate and crisis. Discoveries in geology and archaeology were thrusting humanity deep into the past, thereby complicating traditional biblical timelines. Meanwhile, Darwin's *Origin of Species* (1859) had appeared just six years before *Ecce Homo*, and became a symbol for the naturalistic world picture that was being taken up by a younger generation of men of science hoping to transform the Oxbridge-dominated scientific culture that was epitomized by natural theology. But while heated debates certainly occurred in light of Darwin's great work, we now know that its role in engendering a "crisis of faith" among believers has been largely exaggerated. Many who found the evidence for humanity's ancient and natural origins convincing appropriated versions of evolutionary theory more conducive to their particular religious beliefs.[4] What was truly controversial, however, was the attempted application of a scientific methodology to the study of the Bible and the early Christian past, a method gaining much prominence in the 1860s that was variously called higher criticism or critical philology or even simply biblical criticism. As the late John Burrow remarked in regard to the mid-nineteenth century, "the application of German critical 'philology' to the Bible probably still outranked geology or the theory of evolution as the chief cause of religious doubt."[5] And if not necessarily religious doubt, the critical analysis of Scripture and the early Christian past did cause much handwringing from within the ranks of the English Church, more so than anything offered by the evidence being presented in the natural sciences. And *Ecce Homo* appeared to be underpinned by this highly contentious methodology. But it was not the first English book to do so. Indeed, it was preceded by a truly scandalous book: *Essays and Reviews*.

The controversy over the publishing of *Essays and Reviews* completely dwarfed the controversy surrounding Darwin's *Origin of Species*, which had been published a few months earlier.[6] Appearing in 1860, *Essays and Reviews* spawned hundreds of reviews throughout the periodical and newspaper presses, and at least 150 pamphlets, as well as several volumes of collected essays.[7] It also led to trials for heresy and subsequent appeals, engendering a vast literature that debated and summarized the various legal proceedings. It was a lengthy and an exhausting controversy for proponents and opponents alike that only truly subsided when the book ceased being printed in 1869.

Despite having such an innocuous title, the content, in combination with the identities of the essayists, created a highly combustible atmosphere. The book was a collection of seven individually authored essays. Most significant, however, was the fact that six of the seven essays were written by well-known Anglican clergymen with important positions within the Church. While the essays ranged over a great many subjects, from the history of religious thought to contemporary education, what was contentious was the apparent appropriation and defence of historical and biblical criticism that extended throughout the volume. The German higher criticism was considered problematic because Christianity is not merely a religion. As Owen Chadwick argued, "it is a historical religion,"[8] a religion intimately connected with events that can be dated, analysed, and in theory falsified. In this way the German higher criticism posed a real threat to Protestantism, predicated as it is on the word of the Bible. Therefore Benjamin Jowett's much-quoted statement in his essay "On the Interpretation of Scripture," that the Bible should be examined "just like any other book," was a highly contentious notion indeed. The essayists were thus dubbed "Septum contra Christum," the seven against Christ.

Before *Essays and Reviews* was published, there was a general division in the Church of England between high and low factions, that is between conservative and Tractarian (Catholic-oriented) Anglicans on one side and evangelicals on the other. The publication of *Essays and Reviews*, however, confirmed the existence of the "Broad Church," liberal Anglicans.[9] Largely under the leadership of A.P. Stanley and F.D. Maurice, many of them traced an intellectual and theological lineage back to Thomas Arnold and the Oxford Noetics and perhaps also to the poet Samuel Taylor Coleridge.[10] The Broad Church, however, was never as unified as its critics maintained. While liberal Anglicans tended to agree that advances in science and scholarship should be taken into account in order to reform the faith of the Church and expand its membership, there was a diversity of opinions about the role higher criticism should play in reforming Christian doctrine and about how scientific and historical considerations should be discussed with reference to that doctrine. Should science be made to harmonize with religion or vice versa? And, just as important, how should these views be communicated to a wider reading public? In this regard, *Essays and Reviews* actually represented one particularly extreme approach to these issues. But as it was labelled by critics as a Broad Church manifesto, it had the effect of branding a wide range of liberal Anglicans as theologically radical.

It should be noted, however, that while Jowett's claim about reading Scripture like any other book became a rallying cry for opponents, he ultimately argued, following a well-worn liberal Anglican line, that "When interpreted like any other book, by the same rules of evidence and the same canons of criticism, the Bible will still remain unlike any other book."[11] In other words, by embracing new critical research methods and following the logic of evidence, Jowett believed that Christianity would be purged of its false doctrines, thereby making apparent what was true and universal. This applied as much to the critical examination of the Bible as it did to the study of Christianity's early history. It was this argument that ran through the entire volume, despite the absence of a narrowly defined editorial direction, that suggested "a similarity of purpose and theological perspective" that came to represent the Broad Church party of Anglicanism.[12]

If *Essays and Reviews* announced the unofficial inauguration of the Broad Church, it also united many Anglicans who previously had normally been in opposition to one another. To both conservatives and evangelicals, for instance, *Essays and Reviews* was extremely dangerous, not only because of the rationalist methodology but also, more so, because it was being promoted by Anglican clergymen who were sworn to uphold Anglican doctrines. At this time theological authority in religious publishing was intimately connected with the perceived moral standing of the author, which is an important point worth keeping in mind when we turn to an analysis of the reception of *Ecce Homo*. It was therefore not surprising that many of the angry reviews of *Essays and Reviews* that appeared in the periodical press focused precisely on the identities of the seven authors. Had the essays been written by laymen, it is difficult to imagine such a controversy ensuing. Indeed, the essays might have been ignored entirely. Of course, some of the essays were deemed more reprehensible than others. And, initially at least, the strategy of critics was to divide the essayists among themselves.

Touching off the controversy over the book came from an unlikely source, however; namely, the radical *Westminster Review*.[13] Written anonymously by the positivist Frederic Harrison, the review argued that while the essayists were largely justified in subjecting the Bible and Christianity to new historical criticism, they simply did not go far enough. The logic of their essays did not suggest that Anglican doctrines ought to be salvaged, even in an altered state, but utterly rejected. He found it particularly problematic that their method seemed to entail the discussion of doctrinal problems while offering no solutions to them, apart from vague claims that such critical work would strengthen Christianity in the long run. This view, that the essays only offered a "negative theology," was one that clearly stuck.

Perhaps more significant, however, was what Harrison had to say about the ambiguous nature of the volume's authorship. While it is, effectively, an edited collection, no editor is named on the title page. Moreover, the book opens with a short preface of just three sentences, the first two of which state that "It will be readily understood that the Authors of the ensuing Essays are responsible for their respective articles only. They have written in entire independence of each other, and without concert or comparison."[14] Harrison argued that this was terribly misleading about the essays that followed, as the book itself was clearly a "manifesto from a body of kindred or associated figures" that presented in outline form "the principles of a new school of English theology." Moreover, to say that the authors were only responsible for their own essays was dishonest, as "there is a virtual unity in the purpose of the whole ... and each writer receives a weight and an authority from all the rest of his associates." According to Harrison, "It would be idle to pretend that each writer is not morally responsible for the general tendency of the whole."[15] By suggesting that each essayist was responsible for the views that could be discerned by considering the entire volume, Harrison raised the stakes that were involved in the future reception of the book. Moreover, he challenged the bishops and other Anglican authorities here referred to as "toothless watchdogs of orthodoxy" to make some sort of response. "These professors, tutors, principals, and masters still hold their chairs and retain their influence," Harrison pointed out. Meanwhile, "No authorised rebuke has been put forward."[16]

Unsurprisingly, it was the Bishop of Oxford, Samuel Wilberforce, who took up Harrison's challenge, while making it abundantly clear just how high the stakes had become. At the time of the *Essays and*

Reviews controversy, Wilberforce was also inserting himself into the debates about Darwin's newly published *Origin of Species*. He famously criticized Darwin's theory during the discussion period of a session at the British Association for the Advancement of Science in July 1860 at Oxford University in front of a large audience. His intervention in Oxford foreshadowed his review of the *Origin* that appeared in the *Quarterly Review*, and challenged Darwin's theory as not a science but an ideology that would seek to replace Christianity.[17] This hotly debated review was followed just a few months later by Wilberforce's critique of *Essays and Reviews*, which also appeared in the Tory *Quarterly*.

In the review Wilberforce took Harrison's claim about the unity of the volume, and the related moral responsibility of each of the essayists for the volume as a whole, as a point of departure. Knowing full well that at least two of the essays were largely inoffensive, Wilberforce called on those writers to pull them from future printings. This was directed particularly at Frederick Temple, who was the headmaster of the Rugby school for boys and deemed the most conservative of the group, with the most to lose. Wilberforce openly wondered about the kind of Christian education the well-to-do young men at Rugby could possibly be getting, given the poor judgment Temple showed in attaching his name to such radical and anti-Christian views. "[W]e should tremble," argued Wilberforce, "not for the faith, but for the morals of his pupils."[18] It is clear that Wilberforce used Temple's identity as a respected liberal Anglican to divide the essayists and cause dissension within their ranks. Temple, however, refused to waver. He was subsequently subjected to an embarrassing review of his headmastership by the Rugby trustees.

Though Temple's position was eventually deemed secure, the fate of the other essayists was much less clear.[19] Wilberforce argued that several of them had clearly contradicted specific Anglican doctrines that were formerly specified in the Thirty-Nine Articles of Faith. This was a particularly pointed accusation that was made precisely because it foreshadowed possible legal consequences. The essayists, with the exception of the single layman, had sworn to uphold the Anglican articles upon entering Holy Orders. Wilberforce's suggestion, that in writing *Essays and Reviews* the essayists had possibly committed heresy, opened the door for a much more serious response from the Church of England, one that had clear legal ramifications for the authors of the book.

Making good on the threat, Wilberforce was the driving force behind the drafting of what came to be known as the "Episcopal Manifesto" that was published as a letter to the editor of *The Times* a month after

Wilberforce's *Quarterly* review appeared.[20] Signed by the Archbishop of Canterbury along with twenty-five bishops, the letter was written ostensibly to W.R. Fremantle, a clergyman from Oxfordshire, who had inquired about the Church's response to *Essays and Reviews*. The short letter explains that the bishops had met with the archbishop to discuss the matter, asserting that "They unanimously agree with me [the Archbishop of Canterbury] in expressing the pain it has given them that any clergyman of our Church should have published such opinions as those concerning which you have addressed us. We cannot understand how their opinions can be held consistently with an honest subscription to the formularies of our Church, with many of the fundamental doctrines of which they appear to us essentially at variance." The letter went on to suggest that an appropriate response was being thoroughly considered: "Whether the language in which these views are expressed is such as to make their publication an act which could be visited in the ecclesiastical courts, or to justify the synodical condemnation of the book which contains them is still under our gravest consideration."[21] While the letter indicates that there was still some uncertainty as to what form the Church's response should take, this was the first public indication that legal action was being seriously contemplated. And it was eventually determined that two of the essayists, H.B. Wilson and Rowland Williams, were to be tried for heresy for specific offending passages in their essays. A guilty verdict would mean that that the men would lose their positions in the English Church.

Given that two of the essayists were subjected to trials for heresy, a process that took several years, the controversy surrounding *Essays and Reviews* was a lengthy one. This also meant that, as a commercial endeavour, the book did very well indeed. By January 1862 it had gone through ten editions and sold more than ten thousand copies. To put that figure into perspective, it took Darwin's *Origin of Species*, which was a "moderately successful book" according to Bernard Lightman, ten years to reach the ten thousand mark.[22] Moreover, *Essays and Reviews* continued to sell, reaching well over twenty thousand copies by the end of the decade. *Essays and Reviews* was therefore clearly reaching a large audience of England's reading public. And, despite the strong reaction from the orthodox, along with the very real threat of persecution, *Essays and Reviews* influenced the way in which others approached the Bible. Bishop Colenso is a particularly important case in point in this regard.

When J.W. Colenso first read *Essays and Reviews*, he was the Bishop of Natal, committed to converting the local Zulu Africans to Anglicanism.

And part of his missionary endeavour involved translating the Scriptures into the Zulu language. Aided by "intelligent natives," Colenso argued that in translating the Scriptures he was confronted by objections made by his helpers about some of the historical facts of the Old Testament, facts that were simply difficult to explain. While he initially relied on "the specious explanations" that are normally given to counter "the ordinary objections against the historical character of the early portions of the Old Testament," when it came to translating the story of the Flood, he found himself agreeing with the objections of a particular "simple-minded, but intelligent, native."[23] For instance, Colenso knew that, by the facts of modern geology, the universal deluge of the Bible simply could not have happened, at least not in the way it is there described: "I was thus driven, – against my will at first ... – to search more deeply into these questions .... And now I tremble at the result of my enquiries, – rather, I should do so, were it not that I believe firmly in a God of Righteousness and Truth and Love, who both 'IS, and is a rewarder of them that diligently seek him.'"[24]

The result was Colenso's *The Pentateuch and Book of Joshua Critically Examined*, a book that was researched while Colenso was reading *Essays and Reviews* and its various critical responses. Its publication in 1862 coincided directly with the tenth edition of *Essays and Reviews* along with the continuing trials of Wilson and Williams. Colenso, of course, recognized that the direction of his research aligned him with the *Essays and Reviews* essayists. He therefore found it unfair to suggest that following the apparent consequences of their inquiries would necessarily lead to infidelity or atheism. "Our duty," Colenso argued, "is to follow the Truth, wherever it leads us, and to leave the consequences in the hands of God."[25] With that in mind, Colenso sought to direct his mathematical skills along with the methods of German biblical criticism to a formal study of the Pentateuch.

Colenso was particularly concerned with "the unhistorical character of very considerable portions of the Mosaic narrative,"[26] and he hoped to be able to separate the historically possible elements of the narrative from the downright impossible. Relying on a combination of simple arithmetic and demographic probabilities, Colenso subjected much of the data provided by the Pentateuch to a rigorous analysis. So, for instance, he argued that it was simply impossible for the six hundred thousand men of Israel to have fitted inside the tabernacle in the wilderness, as is claimed in Leviticus 8:14. Colenso also found the stated number of Israelite males at that time completely inconceivable if they had descended from only four generations.[27] Colenso's *Pentateuch* was filled with such painfully detailed accounts of historical and factual inaccuracies. He was therefore forced to conclude that "the main result of my examination of the Pentateuch, – viz. that the narrative, whatever may be its value and meaning, cannot be regarded as historically true, – is not … a doubtful matter of speculation at all; it is a simply question of *facts*."[28]

Colenso was convinced that these factual inaccuracies could no longer be ignored by the Church of England. But he argued that the Church, instead of dealing with these very significant problems, exposed its great "internal weakness" by attempting instead to restrain scientific inquiry.[29] He worried especially that a growing proportion of the educated English population were "in danger of drifting into irreligion and practical atheism" because of the clear disconnect between its spiritual leaders and the views of a great majority of the educated classes.[30] The way forward, for Colenso, was clear: "we shall not rest until the system of our Church be reformed, and her boundaries at the same time enlarged, to make her what a National Church should be, the mother of spiritual life to all within the realm, embracing, as far as possible, all the piety, and learning, and earnestness, and goodness, of the nation. Then, at last, would a stop be put to that internecine war between the servants of one God and the professed followers of the same religion, which now is a reproach to our Christian name, and seriously impedes the progress of truth and charity, both at home and abroad."[31] This was ultimately a message that accorded very well with the liberal Anglicanism of the Broad Church and by extension the Broad Church manifesto that was *Essays and Reviews*.

And, like *Essays and Reviews*, Colenso's *Pentateuch* sold astonishingly well. In just three weeks it went through four editions and sold eight thousand copies.[32] It, too, spawned a large body of scholarship of critical engagement that was, for the most part, quite negative. It was often linked explicitly with *Essays and Reviews*, and by 1864 the connection had become commonplace, no doubt because similar legal actions were being taken against both the essayists and Colenso. On 9 February 1863, for instance, Colenso was sent a letter advising that he resign his bishopric, given that his *Pentateuch* could "not be reconciled with the promises of the ordinal." The letter was signed by every Anglican bishop, with the exception of Connop Thirlwall.[33] Colenso had no intention of resigning. But his metropolitan, the Bishop of Capetown, Robert Gray, who was in consultation with several of those active in the prosecution against *Essays and Reviews*, summoned Colenso to a meeting on 17 November "to answer certain charges of false, strange, and erroneous doctrine and teaching."[34] This was done with a view eventually to bring charges of heresy against Colenso. Colenso denied Gray's jurisdiction, leading to a lengthy impasse that involved Colenso appealing to the Privy Council and an eventual schism in the South African Synod that lasted for decades.

Meanwhile, the charges of heresy brought against Wilson and Williams were initially successful. The two men were suspended from their positions without pay. Like Colenso, they too appealed to the secular courts, where the charges were eventually overturned. The apparently offending passages were deemed ambiguous enough to avoid being contradicted by the Thirty-Nine Articles. In response, the bishops decided to vote on a blanket condemnation of the book at Convocation, a body that included almost twelve thousand clergymen. The motion passed and condemned *Essays and Reviews* "as containing teaching contrary to the doctrine received by the United Church of England and Ireland, in common with the whole Catholic Church of Christ."[35]

What was at issue for both Colenso and the essayists was that they were exposing real factual and historical problems within Anglican theology. What Wilberforce and other orthodox men advocated in response was for the faithful simply to ignore those problems, which, they argued, were not in fact real but rather better understood as mysteries that test the very limits of human comprehension.[36] Moreover, the legal actions that were taken against Colenso and the essayists showed clearly what awaited those members of the Church who sought to subject the "mysteries" of the Bible to a critical analysis.

Meanwhile, the liberal Anglicans, supposedly unified behind their new manifesto, were actually much more divided than the orthodox response to Colenso and the essayists might have suggested.[37] Matthew Arnold, for instance, was particularly scathing in his treatment of Colenso. Colenso may have been factually correct when he challenged many of the supposed historical truths of the Pentateuch, argued Arnold, but he failed to make those corrections in a manner that allowed them to be harmonized with Christian belief. This was also a problem, Arnold noted, with reference to *Essays and Reviews*. "There is," according to Arnold, "truth of science and truth of religion: truth of science does not become truth of religion until it is made to harmonise with it. Applied as the laws of nature are applied in the 'Essays and Reviews,' applied as arithmetical calculations are applied in the Bishop of Natal's work, truths of science, even supposing them to be such, lose their truth, and the utterer of them is not a 'fearless speaker of truth,' but, at best, a blunderer."[38] It was one thing, in other words, for Colenso and the essayists to rely on science and arithmetic to challenge the historical basis for some of the facts of the Bible; it was quite another to do so without harmonizing their critiques with the larger Christian truths that were now being thrown into doubt.

Arnold's comments show that the fracturing of liberal Anglicanism was not just about attitudes to higher criticism but also about publishing strategies.[39] And this was apparent well before *Essays and Reviews* appeared. J. Llewelyn Davies, who was a disciple of Maurice's, was very critical, for instance, of Jowett's earlier attempts to historicize St Paul's epistles by relying on modern rationalist methods of analysis. Davies argued that Jowett's excessive rationalism necessarily erected a barrier between the past and the present, meaning that the eternal aspects of Christianity were largely ignored in favour of those of a historical nature. Too often Jowett appeared to be relying on the "rhetoric and conventional phrases of a school," rather than seeking to get at the eternal Christian truth.[40] This was particularly problematic because of Jowett's "indeterminate" use of the pronoun "we," which had the effect of attributing views to his readers, who may not have been as willing as Jowett to follow the logic of his rationalist exercise.[41] There was, in other words, something underhanded in the way Jowett sought to project his own heavily circumscribed views onto his unsuspecting readers.

This concern for general readers was also central in dividing liberal Anglicans about what to make of Colenso's studies of the Pentateuch and *Essays and Reviews*. Indeed, many believed that Colenso and the essayists failed to appreciate the effect their investigations might have on the wider reading public. The "multitude," according to Arnold, does not inhabit an abstract world of speculation but rather lives in a religious world. When addressing the inhabitants of this world, theological authors especially are under the obligation not to undermine its central truths, particularly by focusing on minor factual inaccuracies that are then used as illustrations to cast doubt on those larger truths. Any new ideas need therefore to be situated within the reality of this religious life. In comparing Colenso with Stanley, Arnold argued that Stanley is much more capable of setting new ideas in an appropriate religious life. He speaks of the Bible "So as to maintain the sense of divine virtue of the Bible unimpaired."[42] "Everywhere he keeps in mind the purpose for which the religious life seeks the Bible – to be enlarged and strengthened, not to be straitened and perplexed." This is what Colenso does not do. By engaging with the Bible on the level of numbers, he has ultimately focused on a largely irrelevant part of it: "the heart of the Bible is not there."[43] Arnold censured Colenso's book because "it puts the non-essential part of the Bible so prominent, and the essential so much in the background, and ... instead of serving the religious life, it confounds it."[44]

Of course, there were plenty of responses in defence of Colenso and the essayists, not necessarily of their views but of their ability to present those views without the threat of persecution. Stanley, for his part, writing for the *Edinburgh Review*, found much to criticize in Colenso's handling of the Pentateuch. He moreover lamented the negative theology that seemed to penetrate all of the essays contained in *Essays and Reviews*. As Arnold made so clear, Stanley had a very different strategy when presenting a historicized conception of Christianity to the reading public. But Stanley found the response of the bishops, particularly their letter to *The Times*, which he dubbed the "Episcopal Manifesto," along with the pending legal action against the essayists, entirely reprehensible. The arguments expressed in Colenso's *Pentateuch* and in *Essays and Reviews* were to be debated, not condemned.[45]

This view was perhaps most eloquently expressed in the immediate aftermath of the "Episcopal Manifesto," also in the form of a letter to the editor of *The Times*. Signed by "A Cambridge Graduate," the letter argued that what was wanted was "not a condemnation, but a refutation." It was no longer sufficient for "ecclesiastical censures" to silence such controversial theological investigations. There was now, according to the letter's author, a large proportion of the laity who were "competent to hear and decide on theological arguments." This audience simply would not be "terrified by a deduction of awful consequences from the new speculations" that were on offer.

This educated audience, the graduate continued, has been taught by philosophy and history alike to seek out the truth no matter how uncomfortable that search might be. For them, the response from the Church was simply unacceptable. "What has hitherto appeared – a couple of intemperate articles in Pharisaical organs; a pamphlet by one of the washiest of High Church bookmakers; an article in the Quarterly, with the usual irritability, and more than the usual unfairness of that review – such things as these are calculated only to alienate the men I speak of." This was the real concern for the "Cambridge Graduate," that intelligent men, who cling in their hearts to Christianity, will eventually leave the English Church because of its inability to confront the kind of problems being exposed by the essayists. "If they can be met and refuted on their own ground, the publication of the book will have been a blessing to the Church; for we cannot ignore the fact that the thoughts they have expressed have long been floating vaguely through the minds of many. The way in which they have hitherto been handled will increase their influence, I think, upon the mass of English laity; it will increase their influence, I am sure, upon the youth of England."[46]

This particular "Cambridge Graduate" was the future Cambridge professor of moral philosophy Henry Sidgwick. That he was expressing the views of a particular milieu is not in question. In this regard, his identifying as a Cambridge Graduate is significant. He was, no doubt, one of the "intelligent men" that he referred to in his letter. He began his education at Rugby, where he learned the importance of responsibility and duty. While at Cambridge as an undergraduate in the late 1850s and early 1860s, he was, according to Sheldon Rothblatt, "completely absorbed in the writings of Mill, Comte, Spencer, Strauss, Renan, Carlyle, Matthew Arnold, George Eliot, and Darwin, wandering freely from biological science to biblical scholarship, ethics and problems of proof."[47] He was excited by the challenges engendered by these thinkers but he also became depressed at the direction such views often led, away from the certainty of his Christian education at Rugby and towards a more uncertain world of biological necessity and moral ambiguity.

When Sidgwick was offered a fellowship at Trinity College after graduating, he struggled with the decision. In order to accept the fellowship, Sidgwick would have had to take Holy Orders, which was a condition of most fellowships at both Cambridge and Oxford at the time. As we have seen with Colenso and the authors of *Essays and Reviews*, there could be serious consequences for failing to live up to the doctrinal conditions that went along with taking Holy Orders. Moreover, it was legally very difficult, if not impossible, to take on a different profession once entering the ministry of the Church. In other words, taking up such a fellowship only to withdraw from it later for religious reasons could be very damaging for one's future prospects. This happened most memorably to the literary critic and historian Leslie Stephen, who found himself unable to continue his duties as a clergyman at Trinity Hall owing to the challenges that recent scientific discoveries were posing to his faith. The existential anguish was apparently so great that Stephen contemplated suicide before giving up his Cambridge fellowship.[48] While Stephen was lucky enough to make a living as a historian and essayist, the prospects for most others in a similar situation would be bleak indeed. Sidgwick was aware of these risks and, given his own recent doubts about certain Anglican doctrines, he struggled mightily before accepting the fellowship and taking Holy Orders in 1859.

In reminiscing late in life about this particular moment in time when *Essays and Reviews* was causing such controversy, Sidgwick argued that he was part of a group of Cambridge scholars whose views ranged "from pure positivism to the 'NeoChristianity' of the Essayists and Reviewers." He claimed that his own theological opinions "were unsettled and widely fluctuating." But, he argued, "[w]hat was fixed and unalterable and accepted by us all was the necessity and duty of examining the evidence for historical Christianity with strict scientific impartiality; placing ourselves as far as possible outside traditional sentiments and opinions, and endeavouring to weigh the *pros* and *cons* on all theological questions as a duly instructed rational being from another planet – or let us say from China – would naturally weigh them."[49] These views were very much reflected in Sidgwick's letter to *The Times*. It is significant that he signed that letter as "A Cambridge Graduate," thereby making it clear that he was not alone.

John Robert Seeley, born in 1834, was also a Cambridge graduate. He entered Christ's College in 1852 and, like Sidgwick, was elected to a fellowship a year earlier, in 1858. Unlike Sidgwick, however, Seeley did not have to take Holy Orders. Christ's College had a long tradition as a reforming institution and had previously established five lay fellowships. Seeley was elected to one of the lay fellowships and therefore did not have to engage in the same sort of existential dilemma as did Sidgwick in making his decision to accept it. There is no doubt that, were Seeley in the same position as Sidgwick, he would have struggled with the decision. This is because, as Sheldon Rothblatt has shown, "All of the problems which absorbed Sidgwick throughout the 1860s wholly occupied Seeley as well and appeared in his first book, *Ecce Homo*, published anonymously by Macmillan in 1865."[50] In many ways, *Ecce Homo* was just the sort of answer to *Essays and Reviews* that many were looking for. It combined a search for truth at all costs with an attempt to illustrate just what united Christians, whether of the Broad Church or High Church, Anglican or dissenting, Protestant or Catholic. Seeley sought to put aside what Leslie Howsam has called the "divisive discourse of doctrine and denomination"[51] and see just what Christ actually stood for from a historical perspective. He clearly did not want to problematize the religious life, to use Arnold's language, but rather engage in the important task of harmonization that seemed to be beyond the reach of the essayists and Colenso. As Maurice Cowling has argued, Seeley "was certainly doing, with different twists and in a different manner, the work which, as Arnold saw it, needed to be done."[52] Moreover, Seeley also hoped that he could give those intelligent men like Sidgwick, who might be straying from the Christian fold, a reason to stay in the Church and try to reform it for the benefit of truth and humanity more broadly.

That this seemingly innocuous message would need to be conveyed anonymously is on first glance difficult to comprehend. Surely all Christians, orthodox or liberal, would have appreciated the attempt to bring to light all that is true and good about Christianity by focusing on the figure of Christ himself. But if biblical criticism was an inherently contentious approach, the historical Jesus was a genre that was equally if not more contentious in part because it was often produced by relying on the same sorts of methods. For all the controversy wrought by *Essays and Reviews* and Colenso's study of the Pentateuch, the central figure of Christianity was rarely discussed. The great fear, of course, was that the appropriation of German biblical criticism would eventually lead to the English version of David Strauss's *Das Leben Jesu* (1835–6). We will see that while Seeley was careful to avoid those particularly troublesome theological areas that Strauss was all too happy to indulge, he was still of the opinion that he had to approach the life of Jesus as an empirical investigator interested in uncovering what the facts themselves suggested. In this way, even though he was trying to provide the positive theology that was missing in *Essays and Reviews*, he was guided by a desire above all for truth and establishing the appropriate method for achieving it. So while there may have been no real legal concerns for Seeley in having *Ecce Homo* published, he was well aware that he was entering into a highly contentious field of inquiry that placed much stress on the identity of the investigator. He no doubt wanted to avoid becoming the centre of a religious controversy. Moreover, it is also clear that he did not want the Seeley name to become associated with heterodoxy or even religious doubt. There were, indeed, many reasons for Seeley to keep his secret.

Chapter Two

By the Author of *Essays on the Church*

Prophets and apostles, from the beginning of the Bible to the end of it, have constantly told us of a coming destruction, of a fiery visitation, of a burning up of the earth.

– *Is the Bible True? By the Author of "Essays on the Church"* (1863)

Before John Robert Seeley was eventually discovered as the author of *Ecce Homo*, the Seeley name was more widely associated with the party of the Church of England that felt most abused by *Ecce Homo*, namely the Low Church or evangelicals. The evangelical party was an already large and growing segment of the Church of England by the time *Ecce Homo* was published. Because of that, evangelical views were highly influential on the overall picture and direction of the Anglican faith, particularly mid-century when evangelicalism was in its heyday. Evangelicals did not just rely on their sheer numbers to establish their cultural authority, however; they were extremely influential and productive in the publishing world as well. As readers began to increase exponentially throughout the early Victorian period, so too did evangelical publishing ventures. Other Church parties simply could not compete with the mass market reached by the efforts of evangelical publishers, who dominated in all areas of the publishing world, from books and newspapers to pamphlets and periodicals in their weekly, monthly, and quarterly guises.

For evangelicals, there was much concern about the availability of cheap books, particularly in regard to quick advances in the realm of scientific and rational thought. Aileen Fyfe has shown, however, that nineteenth-century British evangelicals were not necessarily "concerned about the specific scientific discoveries themselves; rather, they were worried about what they regarded as the distorting manner in which those discoveries were represented to a wider reading public." Major evangelical publishing initiatives, such as the Religious Tract Society, were established to counteract these heretical views while presenting advances in knowledge, whether scientific or historical, as complementary to evangelical Christianity.[1] Other evangelical publishers also did their part to confront this problem, not by denouncing such advances in knowledge but rather by producing their own cheap volumes that explained how new discoveries actually supported rather than undermined the evangelical faith in the word of Scripture. One such publisher was Seeley's father, Robert Benton Seeley.

R.B. Seeley was born into the evangelical publishing business.[2] His father, Leonard Benton Seeley, was a London bookseller and publisher, producing a variety of different kinds of literary works for his evangelical readers while most notably acting as agent for the British and Foreign Bible Society. From his London bookshop, Leonard sold the society's Bibles and New Testaments, thereby contributing to its larger project of distributing the Bible to as many people as possible.[3] R.B. Seeley began working for his father when he was still a child and took over the business in 1826 in partnership with William Burnside. They soon after moved the business from its original Crane Court shop to one on Fleet Street.

Like his father's, R.B. Seeley's business produced a wide range of publications to meet the literary interests and apparent theological needs of his evangelical brethren. One such was the *Christian Lady's Magazine* (1834–49), edited by Charlotte Elizabeth Tonna. It published a variety of different kinds of literary pieces, from poetry and articles to tales from Scripture and theological essays. Most of the articles were unsigned and much was written by Tonna herself. The magazine therefore reflected Tonna's ardent premillennial Tory evangelicalism, a set of beliefs that also conformed to that of the magazine's publisher.[4] Seeley also founded and published the *Churchman's Monthly Review*. A mouthpiece for the Christian Influence Society, the *Churchman's Monthly Review* was conservative in politics and outspoken in its promotion of premillennial evangelicalism.[5] Seeley was also one of a dozen London publishers, including such luminaries as William Longman, John Murray, and John Rivington, to found the *Publishers' Circular*, a periodical that advertised and catalogued the books being published in London.[6]

As a good evangelical, R.B. Seeley's religious beliefs were central to his entire life. His faith encompassed everything.[7] In this regard, he was a leading voice in some of the evangelical social and political controversies of his day. He was an outspoken member of the Christian Influence Society, formed in 1832, which set out to influence the government on religious matters. He also helped to found, along with Lord Ashley (later Lord Shaftesbury), the Church Pastoral Aid Society in 1836, which aimed to help churches deal with the social changes wrought by industrialization in order to bring the Gospel to the new urban poor. According to the first meeting of the Society, it was founded "for the purpose of benefiting the population of our own country by increasing the number of working clergymen in the Church of England, and encouraging the appointment of pious and discreet laymen as helpers to the clergy in duties not ministerial."[8] He also helped found the Society for Improving the Condition of the Labouring Classes in 1844.

R.B. Seeley himself wrote several articles and books that put forward and defended an evangelical understanding of the Christian faith. And, interestingly, he rarely published in his own name. His most popular book, *Essays on the Church*, was first published in 1833. By 1840 it had gone through seven substantially edited versions.[9] The book originated as a response to a dissenting pamphlet that argued that an established national church was unchristian and contradicted Scripture.[10] Seeley sought to counter these claims by providing a "brief and popular, yet comprehensive manual, on the leading points in dispute between the Church of England and the dissenters." He ultimately produced a fairly convincing analysis that established the scriptural basis for the Church of England, one that was hailed by evangelicals and High Churchmen alike. While it was published anonymously, reviewers seemed to grasp, or perhaps had even been informed by the publisher, that it was written by a layman.[11]

In 1834, *Essays on the Church* was extensively revised and enlarged and now explicitly stated that it was written "by a layman." It went through several other new editions in the 1830s, as Seeley continued to update and revise the analysis to take on new challenges to the established Church. In 1838, for instance, the new edition of *Essays on the Church* was devoted to critiquing the rise of the Oxford movement that was led by such figures as John Henry Newman, E.B. Pusey, and the late (as of 1836) Richard Hurrell Froude. Seeley now saw the "modified Popery" of the Tractarians as the main threat to the Anglican faith. This led to a response by Pusey in the form of a 239-page "letter" to the Bishop of Oxford, Richard Bagot.[12] Seeley then produced yet another edition of *Essays on the Church* in 1840 that was almost entirely "re-composed" to meet many of the challenges made by Pusey. The new edition was also prefaced by Seeley's own letter to the Bishop of Oxford, which detailed the main points of his response. In particular, Seeley took issue with the way Pusey designated evangelicals as being "*Ultra*-Protestants."[13] He argued that this was a complete misrepresentation, as he himself, along with most other evangelicals, would never suggest going *beyond* the Reformation, which is what *ultra* must surely mean in this context. Seeley went on to suggest that what Pusey really despised was the Reformation itself, and one need not look further than the contentious *Remains* of Froude to find very clear references to "that odious Protestantism."[14] In other words, it was not ultra-Protestantism at all that Pusey disdained, but Protestantism itself. "The grand question, my Lord, is, which of the contending parties is really in harmony with the Reformation and with the Church? Now their *doctrines*, as well as their *sympathies*, show that it is not the party of the Oxford tracts."[15] This edition of *Essays on the Church* made it clear that it was the evangelicals, rather than the Tractarians, who were the real defenders of the Protestant Reformation, and by extension, the established Church in England.

As each new edition of *Essays on the Church* seemed to engage with new perceived threats to the Anglican faith, the anonymous "author of *Essays on the Church*" itself became an important authorial identity in these debates. R.B. Seeley proceeded to publish a series of books under that persona, signing *Essays on Romanism* (1839), *The Church of Christ in the Middle Ages: An Historical Sketch* (1845), *Essays on the Bible* (1870), and *England's Training: An Historical Sketch* (1885) with the phrase "by the author of *Essays on the Church*," thereby linking all of these works within a common bibliographic lineage.[16] Even though these books were effectively published anonymously, readers would have understood and expected that they were a series and would follow a similar style, voice, and purpose to the others.[17]

It is unclear, however, just how secret this identity was for Seeley. Even though he did not republish any of these books in his own name, his obituaries rather matter-of-factly refer to them as written by Seeley.[18] Moreover, there is at least one reference, well before his death, to "the admirable *Essays on Romanism*, by R.B. Seeley."[19] And one review of *England's Training* argues that while the title page claims that it was written "'By the author of 'Essays on the Church' … it is no secret that the writer who so describes himself is the venerable senior member of the firm that has for so many years published the C[hurch] M[issionary]

S[ociety] periodicals."[20] Rather than being a secret identity, then, it seems more likely that the "author of *Essays on the Church*" was a persona that the elder Seeley would adopt to engage in the particular religious debates of the day.

It is perhaps counter-intuitive, but in the first half of the nineteenth century, as celebrity authors such as Charles Dickens and Walter Scott were multiplying, and their names were becoming important sources of revenue for publishers, anonymous and pseudonymous publishing was still prominent and made up the majority of the book and journal publishing industries. Considering the long history of book publishing in Britain, however, this should not be surprising, as it was not until the eighteenth century, once restrictions were removed from the publishing industry, that the author's name became differentiated from that of the publisher and authorship itself became a profession.[21] And even after that point, there remained important reasons for an author to hide his or her identity, such as "to avoid persecution for libel, heresy, sedition, or other crimes against the State." But after 1800, according to Robert Patten, the reason anonymous and pseudonymous books dominated the industry was because they were commercially viable, and often more so than books that included the author's signature.[22]

This was because not everyone could be a Dickens.[23] An anonymous or pseudonymous work could therefore save the author the embarrassment of producing a failure when the author's name in itself would not contribute to the book's success. If successful, the author's name could then be revealed in a later edition, as was the case with many subsequently famous Victorian authors. Or, the unknown author of a particular book could itself become an identity, linking subsequent books signed "by the Author of …" to a series. By signing his books not with his name but rather "by the Author of *Essays on the Church*," Seeley was therefore following a familiar book publishing strategy. Seeley did not sign his books in this way because he wanted to remain anonymous. Indeed, as we have seen above, it seems that he was quite well known as the author. Rather, he likely referred to himself as "the Author of *Essays on the Church*" because the original book had been successful; subsequently using that title in the place of an authorial signature was a way of reminding readers of the original work and therefore acted as a ready-made form of advertisement for the new one.[24] At the same time, it was a way of transmuting, as Patten puts it, "identity into bibliography" by repositioning "a single title into the latest edition of a putative series."[25]

Figure 2.1 Two books from R.B. Seeley's *Is the Bible True?* series. These pocket-sized books were priced at 1*s*. sewed, or 1*s*. 6*d*. bound in cloth. © The British Library Board. General Reference Collection DRT Digital Store 3155.a.12; 3155.a.12*.

As we have seen, Seeley originally produced new editions of *Essays on the Church* in order to engage in debates with dissenters in the early 1830s and then Tractarians in the late 1830s and early 1840s. But by the late 1850s and early 1860s, a new threat emerged in the form of a new rationalism, engendered in part by scientific discoveries and advances in geology and archaeology as well as via the influence of German biblical criticism. This new threat, therefore, did not lead Seeley to rewrite *Essays on the Church* again but instead to write a series of books signed "by the Author of *Essays on the Church*," a series that would further be linked by the common main title, *Is the Bible True?* (see figure 2.1). His strategy in these books, which no doubt contradicts many stereotypical interpretations of evangelicalism in this period, particularly of the premillennial variety, was often to show that evangelicals had nothing to fear by embracing advances in knowledge as long as such knowledge was properly interpreted.[26] A case in point was his analysis of the "Testimony of Geology."

Written as a dialogue between "James White," a pseudonym Seeley adopted for the series, further compounding the authorial ambiguity of the book that was authored "by the author of *Essays on the Church*," and an orthodox, though naive, evangelical believer called Edward Owen, *Is the Bible True?* confronted the seemingly contradictory views expressed in the Mosaic narrative about the earth's history in light of the relatively recent findings of geology. In the dialogue, Owen acts as the naive evangelical foil for the seemingly much more reasonable and knowledgeable White, who helps Owen understand how to interpret properly the findings of modern science. Owen therefore assumes that White will want to demolish the geological evidence that suggests that the earth is much more ancient than is supposedly suggested by Scripture in order to uphold the views of young earth theologians such as Bishop Ussher, who famously dated God's initial act of creation to 4004 BC. White clearly surprises Owen when he says that the findings of modern geology are entirely convincing. He says that "we must either shut our eyes and ears to geology altogether, or else we must admit its first and most positive assertion that this earth has a 'vast and unknown antiquity.'"[27] That the earth has a "vast and unknown antiquity" is simply undeniable, an argument that is proven by all of the "chief expounders" of geological science. "And here ... we have the last Archbishop of Canterbury, Dr. Chalmers, Dr. Buckland, Dr. Harris, and Hugh Miller, all concurring with the Lyells and Murchinsons, and the other great teachers of geology in our own day."[28] For Owen, this is a shocking revelation that will surely work to undermine the authority of the Mosaic history and thereby further erode the evangelical faith. White suggests that this need not be the case. To the contrary. The recent geological discoveries, argues White, "which at first startled mankind, turn out to be more in agreement with the language of Scripture, than the more limited and erroneous views which prevailed in the last and previous generations."[29]

R.B. Seeley through White essentially argues that the new geological time scales actually help Christians make better sense of the "Scripture narrative of man's creation."[30] This is apparent with even the very first words of Genesis. "In the beginning," White argues, is most likely meant to refer "to some millions of years before Adam's day" and is therefore "more consonant to the general tone and purport of Scripture, with reference to questions of duration, than the narrow view which was current until the last fifty years." Indeed, it is only when we accept a very short time-frame for the history of the earth that we must "'accommodate' or explain away, many explicit declarations of Scripture."[31]

Having established that the ancient time-frame actually accords better with Scripture than a short one, White explains that the fossil record supports the catastrophist views of Cuvier and Murchison, who purport a series of geological eras separated by massive convulsions leading to the destruction and subsequent creation of life.[32] Geology also tells us, according to White, that the succession of these revolutions indicates a visible progression towards some sort of perfection: "that each new creation has been of a higher and nobler kind than that which preceded it."[33] White is adamant, however, that this does not prove some sort of evolutionary developmentalism. To the contrary, White argues that the theory of developmentalism "is utterly broken up and destroyed by the discoveries of geology."[34] Quoting Hugh Miller, White argues that the fossil record makes it apparent that developmentalism is false, given that there are no transitional forms to be found. Each extinct species "is a witness to the fact of creation,"[35] which is "the first and greatest of miracles."[36]

If the clear progression of the geological record does not indicate some sort of evolutionary development, for White, it does provide supplementary evidence for the time of tribulations that precede the second coming of Christ. "The analogy of the past seems to coincide exactly with the declarations of God's word, and to show it to be exceedingly probable, from a view of the past, as well as certain from the declarations of Scripture, that a moment will arrive 'when the earth and all that is therein will be burnt up;' and that after this tremendous event, 'new heavens and a new earth will appear, wherein righteousness will dwell.'"[37] It is in this way that geologists, along with the prophets and apostles of the Bible, confirm White's/Seeley's premillennialism. "Prophets and apostles, from the beginning of the Bible to the end of it," argues White, "have constantly told us of a coming destruction, of a fiery visitation, of a burning up of the earth." And just like geologists now, they have predicted "the rise of 'a new heaven and a new earth,' like a phoenix from the ashes of the old. Again, therefore, the agreement is full and complete. Geology confirms the declarations of Scripture."[38]

Far from challenging the findings of modern geology, White insists that they provide further proof for the Mosaic narrative.[39]

R.B. Seeley is an interesting case study in the evangelical movement mid-century, one that undermines the stereotype that evangelicals were simply dismissive of new discoveries in science because they contradicted Scripture. Properly understood, Seeley was adamant that new scientific discoveries accorded quite well with a literal reading of the Bible. However, he was less willing to accept the application of what he called the "fashionable" historical and critical analysis of the Bible that was first established in Germany and was now making its way to England. He therefore devoted an entire volume in his *Is the Bible True?* series to the challenge posed by *Essays and Reviews* (see figure 2.2).

The pocket-sized volume opens with James White wondering why Edward Owen hasn't been at church lately. It turns out that Owen has been reading *Essays and Reviews* at the Mechanics' Institute and was so interested by it that he took the book home. "I spent a good part of Sunday over it, instead of going to church; and, do you know, it quite took away my inclination to go to church anymore; for if that book is to be believed, then the Bible itself is not true; and if the Bible is not true, what is the use of going to church?"[40] While White believes that Owen's response to the book is quite logical, he explains that Owen must understand that it has that effect on him because it is an entirely dishonest book. It is dishonest precisely because it was written by clergymen who "are holding benefices on the strength of their supposed belief in the Thirty-nine Articles." Given that they produced a book that could only lead believers away from Christianity, they were being dishonest to the service to which they had enlisted themselves.[41] Moreover, White denies the notion that the essays were written individually and therefore have to be judged accordingly. He argues that "there can be no doubt that each one of the seven knew full well what the other six were doing; and, in fact, that they conspired together."[42] According to White, "They have, between them, produced a book which leads every one who believes it to the conclusion that the Bible is not true; and I will put it to your common sense, whether this was a fit work to be undertaken by clergymen of the Church of England?"[43]

Despite White's claim that *Essays and Reviews* was produced precisely to achieve a crisis of doubt in its many readers, Owen argues that he is convinced by their claims that the Bible was basically the fictitious invention of "some early Jewish writer," citing in particular factual errors in the Mosaic narrative and the impossibility of miracles.

IS THE BIBLE TRUE?

SEVEN DIALOGUES

BETWEEN

JAMES WHITE AND EDWARD OWEN,

CONCERNING THE

"ESSAYS AND REVIEWS."

BY THE

AUTHOR OF "ESSAYS ON THE CHURCH.

SEELEY, JACKSON & HALLIDAY, FLEET STREET.
LONDON. MDCCCLXII.

Figure 2.2 *Is the Bible True? Familiar Dialogues between James White and Edward Owen, concerning the "Essays and Reviews"* (London: Seeley, Jackson, & Halliday, 1862). © The British Library Board. General Reference Collection DRT Digital Store 3155.a.12.

About the former, White admits that "Moses describes man and man's earth to have been formed about, or nearly 6000 years ago. But he distinctly intimates that, *before this*, the earth was not non-existent, but lay in a state of darkness and confusion; lasting, it is not said how long."[44] The problem, therefore, is not a case of needing to read Scripture figuratively but literally. When Scripture is understood "by the simplest and most literal sense … the text will bear it … I can conceive nothing more glorious … than the plain history taken just as it stands."[45] When it is read literally, White argues that there is an "overpowering" beauty about the Mosaic narrative, particularly that of Genesis, that reflects the Divine Spirit of its origin. "A sceptical writer has truly said that 'none but a Jesus could have imagined such a character as that of Jesus,'" argues White; "and I repeat the same thought when I say, that none could have described the six days' work detailed in the first chapter of Genesis, except that Divine Spirit who himself conducted that work."[46]

About the miraculous, White admits that, given the fact that "[m]iracles do form so large a portion of the whole Bible narrative, that to tear these out as fictitious, would be to destroy the whole." White therefore suggests that Baden Powell's essay on miracles and the laws of nature is particularly heinous. Given that Baden Powell denied the very possibility of miracles, which he argued simply contradicted the laws of nature and therefore were impossible, White asks Owen "whether this is a proper and decent thing to do, for any man who writes 'Reverend' before his name, and who styles himself a professor in the University of Oxford?" Aside from the moral problems with Baden Powell's position, White argues that what Baden Powell doesn't seem to grasp is that there are essentially two worlds, one for physical science and one for spirits: "The fact is, that 'physical science' has its world and its laws, and spirit has its world also; and hence he who leaves spirit out of his calculation, acts as foolishly as a man who should attempt to make a long voyage leaving all 'trade-winds' and 'gulf-streams' out of his calculation; and who should fix the day of his return without a thought of fog or hurricanes." Baden Powell is wrong, therefore, because he deals with only one aspect of the natural world, while blinding himself "to the tremendous facts that there is a God, and an evil spirit, and other spiritual existences … all of which lie beyond the sphere of his 'law of nature' and his 'physical causes.'"[47] What White seeks to stress is the point that "the establishment of Christianity in the world presents a moral phenomenon, not to be accounted for on ordinary principles."[48]

Owen is, of course, eventually disabused of his "faith in the Essays in Reviews" but he admits that his belief has been shaken and has not been "restore[d] … to its former strength." He claims to be "now like a man recovering from a serious illness – I have no firmness of tread. I wish that, in addition to getting rid of the *negative*, you could establish me in the *positive*. I see that disbelief of the Bible is not a reasonable thing, but I lack a hearty and robust belief."[49] White, of course, believes that this is precisely the effect that was intended by the *Essays and Reviews* essayists because they did not seek to present a positive theology but merely sought to tear down the edifice. The good news for Owen is that he clearly wants a positive system of belief. But he needs to be convinced that evangelical Christianity is just that system. White suggests that all Owen needs to do in order to be convinced is to examine the state of the world. Its "brightest parts" are those that have been founded on biblical principles, and its most "miserable" are "those parts of the world where no Bible is found."[50] To drive his point home, White argues that "man without the Bible, is in a condition of fearful degradation, darkness, and misery, from which he has no power to raise himself; and hence, that if there be a benevolent Deity – a fact which even sceptics generally admit – there is nothing more probable, nothing more consonant to his nature and character, than that He should stretch forth his hand to help and rescue these his creatures from the wretchedness into which they are continually plunging themselves."[51] White asks Owen again, "whether, as Christianity seemed to wane and lost ground in your estimation, any other systems or mode of belief took its place? Did the writers of 'Essays and Reviews' show unto you a more excellent way?"[52] Of course they did not.

When Colenso's study of the Pentateuch was published shortly thereafter, many of Seeley's criticisms of *Essays and Reviews* were subsequently repeated against Colenso. Owen, once again, seems on the cusp of losing his faith, arguing that he was convinced that either Colenso "was altogether in error, or else the Bible must be given up."[53] Unlike his discussion of geology, White agrees that in this case there is no middle ground. But Colenso's views were not motivated by modern science, argues White. Rather, Colenso is entirely motivated by a fashionable theory, that of Rationalism, in much the same way that several Oxford dons were taken in "twenty years ago" by the fashionable bias of "Popery," that is to say Tractarianism. According to White, "Truth" was never the goal of "either of these hostile camps."[54] Worse still, says White, is that Colenso's work is entirely beholden to the supposed authority of German biblical scholarship, which is by necessity inherently anti-Christian, because of the way in which it "freely and rudely handle[s], as human," that which is "Divine."[55] For White, the authors of German biblical criticism are not interested in examining "Holy Scripture with a mind absolutely free from purpose or bias" but are rather motivated entirely by "fame" and "distinction" and therefore set out to pronounce "bold and novel speculation[s]" about the "unhistorical" nature of Scripture.[56] Such is deemed the case with Colenso as well, who "clearly shows a spirit of carping and hostile criticism, which taints his whole argument."[57]

A particularly telling criticism of White's involves Colenso's claim that there could have been at most five thousand Israelites at the time of Exodus rather than the biblical figures that placed the population somewhere around six hundred thousand. Applying Malthusian rates of population increases, White argues that the biblical figures are actually quite conservative and that on this issue Colenso "[n]ot only fails, but fails disgracefully." Perhaps most frustrating for White is the fact that a naive reader could be led by Colenso's blatant errors of basic calculation to question the truth of the Bible. This is simply unconscionable, and White claims that he is rather struck by a "sort of mingled feeling" of both anger and pity. He is angry "that a man should assail God's Word on such weak and frivolous grounds"; but he also, more importantly, feels pity "when I thought of the guilt connected with such an attack." Colenso, therefore, "must be severely condemned for the flippancy and hollowness of the arguments with which he assails the truth of this portion of God's Word."[58]

White admits that there is a theology that is promoted by Colenso, but it is not that of Christianity. Indeed, much like the *Essays and Reviews* essayists, Colenso entirely undermines the historical claims of the Bible, and therefore its authority as the word of God. What Colenso seeks to engender, therefore, is a Christianity of the heart rather than a Christianity of the book. "You know we talked of this, when we were considering *Essays and Reviews*," argues White. "It simply brings us to the point where Rousseau, and Hume, and Voltaire, and Thomas Paine, all stood. All these men, and all their followers, were quite willing to admit that there was a God, and that there were fine and noble passages in the Bible. But they all claimed, just what Bishop Colenso claims, an absolute right to deal with the Bible as judges and arbiters – to take what they pleased; assigning no other reason than this, that one part commended itself to their hearts, and another part did not. In Bishop

Colenso's view, the heart of man is to be the sole judge of the truth and value of every single passage of the Bible."[59] Colenso's theology, then, is essentially that of Deism, "the most cold, dead, inoperative thing in the world." And much like Deism, Colenso's purpose is to destroy belief in the Bible. The ultimate effect of Colenso's book "is nothing short of soul suicide."[60]

Like many evangelical publishers of his day, Seeley believed that it was his duty to inform evangelicals about contemporary theological controversies while presenting them in the light of his own particular premillennial system of Christianity. He stressed that evangelicals had nothing to fear from scientific advances, as long as they were properly interpreted. This, too, was extended to his understanding of the critical examination of the Bible. In this regard, a proper interpretation was a literal one that accepted its scientific and spiritual truths. The *Essays and Reviews* essayists, along with Colenso, were therefore to be condemned for trying to rationalize that which could not be rationalized while ultimately producing a negative theology that could only convince readers that the Bible was not the true word of God. For Robert Benton Seeley, in other words, the appropriate response to the religious crises of the 1860s was simply to reinforce the authority and the positive theology produced by a literal reading of the Bible. His son, however, had a very different approach in mind.

Chapter Three

Father and Son

Christianity seems to me to teach that the perfect revelation of God is in Christ – not in the Bible, but in Christ.

– John Seeley to Mary Seeley, 3 April 1859

R.B. Seeley was fairly typical of his generation of evangelicals in the sense that he did not question the central tenets of the evangelical faith. Those evangelicals who grew up in the first decades of the nineteenth century were still largely a minority among Protestants and therefore had, as Aileen Fyfe puts it, "great ambitions to influence the religious opinions of the rest of the country, and ultimately the world."[1] Through the efforts of publishers like Seeley and publishing ventures such as the Religious Tract Society, by the 1850s evangelicalism had moved into the religious mainstream, very much realizing many of the goals of the early evangelicals. But this meant that children who grew into maturity during this period of evangelical dominance rebelled against their religious training, often rejecting the evangelical faith entirely.[2] An example of this is detailed in *Father and Son* (1907), Edmund Gosse's famous account of his upbringing within a fairly radical evangelical group known as the Plymouth Brethren.[3] The book describes, in painstaking detail, Gosse's rejection of his father's faith, which is held up for ridicule whenever it seemed to force the older Gosse to accept untenable positions, particularly in regard to his work in natural history. Philip Gosse not only was well known for arguing that God put fossils in the geological record to test our faith but also used similar logic to explain why Adam had a navel.[4] While R.B. Seeley's evangelical beliefs were not quite as extreme as Gosse's, several of his ten children did not wholeheartedly agree with his faith. His third child proved particularly rebellious in this regard.[5]

While he never succumbed to the full-out atheism of unbelievers such as Edmund Gosse, John Robert Seeley came to sympathize with the Broad Church views that his father so clearly condemned in his treatment of *Essays and Reviews* and the work of Colenso (see chapter 2). It would be going too far to suggest that the fictitious and naive Owen from Seeley's *Is the Bible True?* series was a representation of his son, but it is clear that Robert knew well that John's faith was being challenged by the religious controversies of the 1860s and that his sympathies were very much in the direction of the Broad Church. Although very little of their correspondence has survived, two letters written by John sometime in the 1850s, while he was an undergraduate at Christ's College, give wonderful insight into the burgeoning theological differences between father and son as well as the openness with which these differences were expressed. Indeed, John clearly took Robert at his word when he said that there "ought to be no reserve" in their theological discussions.

In the first letter, John attempted to explain further to his "Papa" the theological views of F.D. Maurice that John largely agreed with, though not uncritically. For Maurice, explained John, there is a clear distinction between religion and science, a "distinction equivalent to that of Coleridge between the understanding and reason." Maurice's point, in this regard, was that religious belief cannot stand or fall on the basis of rational tools of scientific verification. For a scientific theory to reach the status of "truth," there must be some consensus. But that is precisely what is missing when it comes to religion. In this way, "religion is unscientific because there is *infinite* disagreement about it." The existence of God is "easily believed and believing it has done much good, but it cannot be proved scientifically nor yet, I think, historically and I am not satisfied that any other kind of proof ought to be received." Theology, therefore, "must be submitted to a different kind of investigation" from that of science. According to Maurice, theology itself must only "be deduced from moral instincts."

One of the specific implications of this view, according to John, had to do with the reading of Scripture, which could now be liberated from any adherence to completely literal interpretations. This the younger Seeley particularly admired about Maurice. And given the elder Seeley's view that the best way to understand Scripture was to read it simply and literally, he certainly would have disagreed with Maurice on this point. John wrote: "I like the kind and degree of respect he [Maurice] shews for the authority of Scripture not regarding every clause, word and letter of it as the word of God … I like the method [of] interpretation he employs trying to discover the meaning of each passage by itself not considering it his province to reconcile discrepancies nor … to construct a system but confident that if he takes one of the parts the whole will take care of itself." And here Seeley explicitly contrasts Maurice with evangelicals: "when I see the Evangelicals straining Scripture I say to myself: 'these men are so wedded to a theory that they really cannot see that which to me is as clear as daylight.'" And what is "clear as daylight" for John is the stress that Maurice places on the "mission of Christ," his "perpetual presence … and his right to the allegiance of men," which is "more the subject of Scripture and the practical part of Christianity than his suffering and death." In this way Seeley was more than happy to agree with Maurice "in exploding the Evangelical notion that there must take place in each man's mind a kind of dramatic representation of the great act of redemption once accomplished for the whole world accompanied with a prescribed series of violent emotions – a notion which proclaimed constantly from pulpits makes many children, I have reason to believe, practical Atheists because the required emotions will not come at call." These must have been difficult words for John's father to read, particularly the suggestion that certain evangelical beliefs necessarily lead certain children down the road to atheism.[6]

Robert's response unfortunately has not survived, but a subsequent letter from John suggests that Robert dismissed Maurice as an unbeliever.[7] Worse, given that Maurice was an ordained minister of the Church of England, Robert must have implied that Maurice was dishonest about his devotion to the Church, given his views about the atonement. (As we saw in the previous chapter, the issue of honesty was central for Robert in regard to the debates about *Essays and Reviews*.) In his controversial *Theological Essays* (1853), Maurice had denied the doctrine of the eternal punishment of unrepentant sinners, questioning in particular the notion that "eternal" was necessarily synonymous with "everlasting." Because his own conception relied upon John 17:3, "This is life eternal, that they should know thee, the one true God, and him whom thou hast sent, even Jesus Christ," Maurice was unwilling to see eternity as having anything to do with duration. "I must see eternity as something altogether out of time." Evangelicals were especially annoyed because of the practical consequences that were foreseen as resulting from the removal of the main sanction for sin.[8] For Robert, in debate with his son, Maurice's denial of atonement meant that he could only have dishonestly subscribed to the Anglican articles, much like the

Essays and Reviews essayists and Bishop Colenso. This was also apparently the view of the Council of King's College, which forced Maurice to resign from his professorial chair. Indeed, because of Maurice's *Theological Essays*, he was often associated with *Essays and Reviews* even though he was himself ambivalent about historical criticism.[9] (And, as we saw in chapter 1, Maurice's disciple J. Llewelyn Davies was quite critical of Jowett's historical criticism.)

Maurice was, therefore, a controversial religious figure. The question of Maurice's honesty and denial of atonement must have been an ongoing debate between father and son, a debate that was growing quite tiresome for John, who was clearly unable to convince Robert that at issue might simply be a difference of interpretation rather than outright dishonesty. "If however," John wrote, "you continue to decline advancing proof and to assume in letter after letter that your interpretation of the Church formularies is for some reason or other better than Maurice's and that by his interpretation the laws of language are violated while by yours they are all satisfied, I suppose I must advance something to the contrary, although I really do not see that I am called upon to do so." In defending Maurice's interpretation, John argued that the word "atonement" had an entirely different meaning "when the Church formularies were made." One can find in Shakespeare, for instance:

> There is joy in heaven
> When earthly things made even
> *Atone* together.[10]

In terms of the practical issue of removing the notion of eternal hellfire, John argued that Maurice's view of atonement is actually a more realistic safeguard against sin: "You will say that it is not an unimportant matter which sense is given to the word eternal whether 'of infinite' or 'of indefinite duration,' that the practical moral force of the doctrine is destroyed, if the sinner can flatter himself that his punishment will have an end. But surely it may be answered that a sinner so flattering himself is doing his best to make his punishment endless. And it may also be advanced that what the doctrine loses at one end it gains at the other, for if evil that ends, has less effect upon the imagination than evil that lasts for ever, so also has evil that is future less than evil that is present." It followed that such a view of atonement was more accurately Christian than the alternative: "while the Evangelicals tell the sinner that after death he will be damned, Maurice tells him with St John that he is damned already."[11]

John engaged in a similar debate a few years later with his sister Mary, who at the time embraced the evangelicalism of her father. John suggested that Mary read the recently published *Unity of Evangelical and Apostolic Teaching* (1859), a volume of sermons by A.P. Stanley. There she would find out more about the "Broad Church Theology" that John was advancing and hopefully understand that "the root of our difference is that we have not the same rule of faith." Rather than seeing "the Bible as the sole revelation of God," John argued that he believed that "God reveals himself in ten thousand ways, by the order of the world, by the relations of parent and child, husband and wife, brother and sister, by history. I therefore believe that God is partially revealed to everyone, that all persons in this world have light enough to walk by, knowledge enough of God to make virtue possible and to prevent the light of the Eternal Life from being extinguished in them." Foreshadowing his future survey of the life and work of Jesus Christ, John told Mary that "Christianity seems to me to teach that the perfect revelation of God is in Christ – not in the Bible, but in Christ." It was through the "character of God" that we should draw conclusions about proper moral and theological instruction, not through intricate and literal readings of Scripture. And when we do this, argued John, in following Maurice on Atonement, we will recognize "that God does not damn people for heterodoxy, no, not for Atheism."[12]

Unlike Edmund Gosse, who was unable to find any persuasive theological framework to replace the views of his father that he so thoroughly rejected, John Seeley embraced what he referred to as Broad Church theology. As we saw in chapter 1, the theological views often falling under this rubric were remarkably diverse, with the divisions of liberal Anglicanism ironically becoming ever more apparent following the publication of *Essays and Reviews* and the works of Bishop Colenso. At the same time, Seeley was clearly putting together his own version of liberal Anglicanism that included the eschatological views of Maurice supplemented by Stanley's belief that God and not the Bible was the key to salvation. Perhaps unfairly, these diverse, liberal theologies became associated in the 1860s with the biblical and historical criticism of *Essays and Reviews* and Colenso's works. R.B. Seeley, for one, believed that the embrace of German historical criticism was the logical extension of his son's Broad Church theology, a "theology" that could only have the effect of making unbelievers out of the faithful, eternal sinners out of the virtuous. When the older Seeley claimed that Colenso's adherence to Deism was essentially soul-suicide, one wonders what he would have thought about his own son's soul. Perhaps some of this helps explain why John Seeley decided to write a book about Jesus Christ. He wanted to show his father and others like him that it was possible to convince those headed towards unbelief to stay within the Anglican fold, not by appealing to the historical facts of the Bible, as was, ironically, both his father's strategy and that of the *Essays and Reviews* essayists along with Bishop Colenso, but by appealing to the person and work of the son of God.

Producing just such a study of the history and life of Jesus Christ was no unproblematic task, however. If biblical criticism, as an approach to the study of theology, was a particularly troublesome minefield, the "historical Jesus," as it came to be known, was an even more controversial genre of religious and theological scholarship.[13] Indeed, the historical Jesus was at the time becoming a contentious subject of both general and academic interest, though there had been no English biography of Christ that could compete with the studies appearing across the channel and subsequently in English translation. Most notable was David Strauss's *Das Leben Jesu* (1835–6), which was a substantial contribution to the historical Jesus genre, not just for its careful historical reconstruction but perhaps more so for its application of the so-called higher criticism to the facts of Christ's life, a method that included a healthy dose of Hegelian idealism as well. For Strauss, it made little sense to try and reconstruct the actual life of Christ from the Gospels, because the Gospels themselves were best understood as myths, as fantastic stories constructed after the events in question in order to fit idealized religious beliefs. The life of Christ that is contained in the Gospels, therefore, is an imaginative creation that was developed out of a mythopoeic process. Jesus was constructed in a particular way to fit prevalent Jewish religious conceptions of the time. It was without debate for Strauss that the miracles and supernatural events that are recounted in the Gospels and that therefore prove Christ's divinity were imagined fictions.[14]

Strauss was initially translated into English in 1842, and by 1846 a respectable translation was produced by Marian Evans (later better known as George Eliot) that was widely noticed. A particularly representative review in the *Westminster and Foreign Quarterly Review* found that, primarily, Strauss "*un-creates*. At his spell," the reviewer continued, "the warmth of every faith, the accumulated glow of old ages, that alone renders the Present habitable, suddenly becomes latent: the facts, the scenes, the truths that re-absorb it, run down in liquefaction, pass off in vapour, and restore the world to a nebular condition."[15] As Jennifer Stevens argues about the larger implications of this review, "the arresting notion of 'un-creation' and the images of deliquescence convey a hauntingly desolate picture of a post-Straussian world, in which civilisation reverts to original chaos."[16] Of course, this would have been the view of R.B. Seeley as well. As he stated in his criticism of *Essays and Reviews*, those societies that do not accept the Bible as the word of God necessarily become degenerated wastelands. *Das Leben Jesu* was simply too radical even for liberal churchmen. At the time the book had the counterintuitive effect of strengthening orthodox interpretations of the life of Jesus in England.[17]

The other significant historical Jesus that appeared in advance of *Ecce Homo* was Ernest Renan's *Vie de Jésus* in 1863, which was translated into English before the end of the year. Renan was on his way to becoming a priest before his reading of the German higher criticism caused him to lose his faith. But while he was capable of the kind of close historical reading of biblical texts typically associated with German authors such as Strauss, he wrote in a captivating narrative style that won him thousands of readers throughout Europe. He was outspoken about his denial of miracles and the supernatural, and on this fairly fundamental level his history of Christ was a thoroughgoing rational one. Unlike Strauss, however, "Renan believed the historian could recover a good deal of authentic data on the life of Christ." According to Daniel Pals, "[o]ne could detect the basic thread of his ministry, his personal and religious motives, something of his travels and confrontations, and certainly the rare spiritual impact he made on those he met."[18] Renan's biography tells the story of how an earnest young man who started out preaching a simple message of piety and love was transformed by those surrounding him, notably by John the Baptist, into a fanatic who encouraged his followers into believing his false miracles. Renan does not discuss in detail the Resurrection, but it is clear that he thinks it was in part the product of the imagination of Mary Magdalene and was therefore not a factual event worth narrating.

Even though *Vie de Jésus* was a book about the life of Christ that took the Gospels seriously as historical documents of factual import, and not therefore as merely evidence for the mythical beliefs of the time, it was largely rejected in the English periodical press. "As with Strauss," according to Pals, "it seemed to be the denial of miracle that offended most."[19] Even those churchmen most likely to sympathize with the critical method employed by Renan, namely those liberal Anglicans so often designated with the Broad Church label, found his approach to be far too radical. Some of the most critical pieces were published by Macmillan and Co., a publishing firm that was becoming the "intellectual centre" of the Broad Church.[20] Such pieces included John Tulloch's pamphlet that found Renan's history to be the "consummation" of positivist thought, "divorced from all faith and true reverence," thereby marking "the spring-tide of an advancing wave of thought inimical to Christianity."[21] In *Macmillan's Magazine*, none other than F.D. Maurice gave an ironic assessment of *Vie de Jésus*. There were lessons to be learned from the book, argued Maurice, despite its basis in the imagination of its author. If Renan's chief (and false!) argument was that "Jesus was born 1800 years ago, [and] that He had no life before, or has had since," this should remind the pious in England about their own tendency "to glorify the past, or the present, or the future, at the expense of the other." What Renan truly denied in his historicization of Christ, therefore, was the Eternal. "[I]f we would be in the true sense humanists, being willing to be denounced as bigots by those who usurp the title," argued Maurice, "we shall speak of a Living Christ – of One who is, and was, and is to come."[22]

Seeley would have been aware of the criticisms directed against the historical Jesus genre, particularly the large, negative reaction to both Strauss's and Renan's denials of supernaturalism along with their scepticism regarding the factual accuracy of the Gospel narratives. But Seeley was also no orthodox churchman looking simply to provide a narrative of Christ's life that would reinforce all of the theological and supernatural truths that someone like his father would likely have wholeheartedly endorsed. Even though he believed himself to be a Christian, and in particular a Christian of the established Church, he was also influenced by the rationalism so hated by his father. Moreover, while his time at Cambridge introduced him to many of the thinkers and proponents associated with the Broad Church, when he wrote *Ecce Homo* he had since completed his fellowship at Christ's College and was a lecturer of ancient history at University College, London. At that secular institution he met many secularists, atheists, and, perhaps more importantly, proponents of positivism. He believed, moreover, that their attempts to create a "science of society" were much more aligned with a form of Broad Church Christianity than his friends from Cambridge might have realized.[23]

Positivism was originally developed by Auguste Comte in the first half of the nineteenth century as a philosophy and a method of inquiry that promoted, above all else, the necessity of scientific understanding of all phenomena.[24] What was particularly novel about Comte's approach was his claim that the same method that was applied to the physical sciences needed to be applied to the study of society as well, and this included human history, politics, and social relations in general. He argued, moreover, that his positivist philosophy was itself a product of a law of social development. It represented the last, positivist, stage of history that was preceded by two others, a theological stage and a metaphysical stage. Society was only just breaking free of a metaphysical stage of knowledge, which was still corrupted by appeals to supernatural causes, and entering into the positivist stage, which, by definition, rejected appeals to a first cause and embraced naturalistic explanations for all phenomena, including those governing human society. The appearance of sociology, in particular, a term invented by Comte, was a profound symbol of this new positivist age.

It is certainly the case that, as positivism gained a hearing in Britain largely through the work of John Stuart Mill and the *Westminster Review* along with selective appropriations by historian Henry Thomas Buckle and evolutionist Herbert Spencer, to name a few, many religious figures rejected it as a form of atheism. However, while Comte explicitly rejected theological and metaphysical explanations of phenomena, he did not necessarily do away with religion. Early critics of Comte argued that his materialism banished God from the universe, but others were able to argue that God was central, given the beauty and order of the system. As David Masson, the future editor of *Macmillan's Magazine*, argued in an early essay on Comte in the *North British Review*, it was possible to read into Comte "the hand of God in history … carried on through the medium of what are called laws."[25] It was therefore possible to reconcile a providential view of historical development along with Comte's claim that all phenomena, physical or human, are the product of orderly laws.

Moreover, Comte himself would come to realize that, in order for humanity to come to terms with the new positivist age, which also included a new industrializing reality, the establishment of a new ethical and moral foundation would be necessary. He then expanded his positivist system to include a "Religion of Humanity," which would be founded on the principle of "altruism," another term that Comte invented to describe the mutual obligations humans would come to feel towards one another within this new positivist society. Many of Comte's English followers did not take this further step with him, as positivism the philosophy was often separated from positivism the religion. This was largely because Comte's Religion of Humanity also came with highly structured liturgy that resembled Catholicism, so much so that Thomas Henry Huxley famously referred to the Religion of Humanity as "Catholicism minus Christianity."[26] And yet, as John Stuart Mill was to illustrate in his popular *Auguste Comte and Positivism* of 1865, Comte's positivist religion was underpinned by ideals that were gaining prominence among Christian socialists at the time, even while they too would have rejected the pseudo-Catholicism of Comte's liturgy.[27]

It was from this perspective that Seeley found himself engaging with the positivists of London. Not only was Seeley a close colleague of one of the main expounders of positivism, Edward Beesley, who was professor of history at University College, London, but also there was much contact between positivists and the Christian socialists Seeley associated with in London. Both groups were finding common cause in improving the welfare of working men and women. One particular initiative included the establishment of the London Working Men's College in 1854, which was founded by Maurice along with a group of Christian socialists. Seeley taught courses there in the early 1860s along with many of Maurice's other admirers. But many positivists taught there as well, most notably Frederic Harrison. The positivists recognized, as did the Christian socialists, that a liberal education was central to improving the lives of working men and women.[28] It is true that Harrison believed that there was no middle ground between the liberal Anglicanism of Maurice and the positivism of Comte, as was most apparent in his review of *Essays and Reviews*. Nor did Maurice find much to praise in positivism at this point either, though he would come to regard it more sympathetically later in the century. But even at this point some of the figures associated with Maurice and the Christian socialists, such as J. Llewelyn Davies as well as Seeley, clearly found something in positivism that appealed to their desire to restructure theology along modern lines.

What particularly appealed to the Christian socialists about positivism was its stress on the universality of humanity, which related quite closely to the Christian idea of a universal brotherhood. But with the divine removed from the positivists' scheme, even some of its Christian socialist sympathizers worried about what would happen to the Christian foundation of morality in a positivist society. Just what was the relationship, therefore, between a positivist focus on the welfare of humanity and Christian morality? Was it possible to integrate with the positivist philosophy of history a story of the founding of the Christian society by focusing on the figure who best represented the altruistic ideal of humanity? These were certainly some of the questions that motivated Seeley's inquiry in the life and work of Jesus Christ.

Ecce Homo, then, was very much driven by a series of seemingly competing methodological, philosophical, and theological impulses that make for a strangely ambiguous narrative analysis of Christ's life. Seeley wanted to show that it was possible to present a critical analysis of Jesus's life, one that relied on the facts alone, while not essentially fictionalizing the New Testament and undermining the Christian faith. In a sense, Seeley wanted to provide just that positive theology that the *Essays and Reviews* essayists were unable to achieve, by showing that the life and philosophy of Jesus Christ could be subjected to a rigorous historical examination that would thereby clarify what was truly central and eternal about Christianity and what was otherwise a product of historical circumstance. *Ecce Homo* would become a profound symbol of the way in which nineteenth-century theories of historical development, biblical criticism, and positivism were often entangled in attempts to reformulate Christian belief and practice in the modern age. For many orthodox believers like Seeley's father, however, any one of these approaches could only weaken Christianity, not strengthen it.

Chapter Four

The Victorian Jesus

Of his two great gifts, the power over nature and the high moral wisdom and ascendency over men, the former might be the more astonishing, but it is the latter which gives him his everlasting dominion.

– *Ecce Homo: A Survey in the Life and Work of Jesus Christ* (1865)

In the next few chapters, it will become clear that *Ecce Homo* meant very different things to different readers. But before examining the diversity of those meanings, and how that diversity related to *Ecce Homo*'s hidden authorship, it will be necessary, at the very least, to summarize the main features of the book itself in order to better grasp just what was being said about it. Most significantly, it must be understood that while it is often referred to as a biography of Christ, this is misleading.[1] The subtitle, of course, has played a role in this confusion. Some reviewers noted the problem and expressed irritation that it was not really "A Survey in the Life and Work of Jesus Christ." In fairness to Seeley, it does give a survey *in* some key aspects of Christ's life and character. But what it does not do is give a detailed narrative of Christ's life, from his birth to his death and resurrection. It is clear that Seeley was not terribly concerned with presenting such a narrative. At one point in *Ecce Homo* he says that because the details of Christ's life are already well known thanks to the Gospels of the New Testament, which he often refers to as "Christ's biographies," it is unnecessary for him to repeat them, except in instances when those details relate specifically to his argument.[2] What the book is, instead, is a carefully constructed argument about the true meaning of Christ's life. It is historical in the sense that Seeley sought to situate Christ and the society he founded within a particular ancient context. But this was done, ultimately, to discern just what was eternal about what Christ thought and did. What, in other words, did Christ have to offer to a modern society that was now two thousand years in advance of the age when he was alive?

Ecce Homo opens with a very short preface – under two full pages – that tries to give a clear and concise explanation of its purpose. Seeley says that he wrote the book because he was "dissatisfied with current conceptions of Christ." He does not say what was so dissatisfying about those current conceptions or mention any particular studies by name. But this dissatisfaction was so great that he says that he was forced "to reconsider the whole subject from the beginning" and place himself "in imagination at the time when he whom we call Christ bore no such name, but was simply … a young man of promise, popular with those who knew him and appearing to enjoy Divine favour." In order to present such a historically accurate view of Christ, Seeley argued that it would be necessary "to trace his biography from point to point, and accept those conclusions about him, not which church doctors or even apostles have sealed with their authority, but which the facts themselves, critically weighed, appear to warrant."[3] He wanted to engage in this study because he found that, even "after reading a good many books on Christ, he felt still constrained to confess that there was no historical character whose motives, objects, and feelings remained so incomprehensible to him."[4] As we will see, readers noticed that there seemed to be a mixed message being relayed in these brief statements. On one hand, it appears that Seeley has embraced the higher criticism of Strauss by claiming to rely on facts that have been "critically weighed." But on the other, he seems to accept Christ's divinity, given that he wants to go back to a time when Christ "appear[ed] to enjoy Divine favour."

By the end of the preface, however, it is apparent that Seeley, for the most part, wants to avoid the main points of controversy that surrounded Renan and Strauss. After stating that it must be understood that *Ecce Homo* is but a "fragment," Seeley goes on to baldly state: "No theological questions whatever are here to be discussed." As we will see in the next chapter, this became a particularly contentious methodological claim. How could the life of Jesus even be conceived without making assumptions about his divinity, never mind engaging with the supernatural facts of his life? This was clearly a reaction that Seeley expected; he hoped that by stating that a future work would deal with Christ "as the creator of modern theology and religion," he would pre-empt any sustained attacks on that front. But this meant that the findings of *Ecce Homo* were often thought to depend on the promise of a future volume, a promise that was left far too long unfulfilled (see chapters 12–13). With that said, Seeley stressed that the current study was a simple one with a rather narrow purpose: "to furnish an answer to the question, What was Christ's object in founding the Society which is called by his name, and how is it adapted to attain that object?"[5]

In answering these two related questions, Seeley divided *Ecce Homo* into two parts. The first was largely an early history of the Christian Church in connection with Christ's life (Christ's Call); and the second, much longer part of the book (to be discussed below) was a meditation on the central Christian principles that Seeley argued derived organically from Christ's life and character (Christ's Legislation). The first part of the book, then, was really an attempt to establish the historical context for the Christian principles that were discussed in the second part.

Seeley began his discussion of the historical context by suggesting that Christ was born into a society that was prepared to receive his message. This was because, as Seeley argued in the first line, "The Christian Church sprang from a movement which was not begun by Christ." Preceding Christ was "the first wave of a movement" that had passed over Judaea and the Roman Empire and which ultimately called for a new morality. "There was a clear stage," Seeley argued, "as it afterwards appeared, for a Universal Church." And before Christ made his appearance, "the whole nation were intent upon the career of one who was attempting in an imperfect manner that which Christ afterwards fully accomplished." This was, of course, John the Baptist.[6]

John was an important precursor, according to Seeley, because he "revived the function of the prophet." While prophecy had long been discontinued, bringing prophecy back into the ancient world was an important step towards the founding of Christianity. This is because prophecy unites in a particularly effective way, Seeley argued, "all that is highest in poetry and most fundamental in political science with what is most practical in philosophy and most inspiring in religion."[7] By establishing this "prophetic authority," which included amassing a large group of followers through his practices of baptism, John essentially took the first steps towards establishing the "Universal Church."[8] It was, of course, beyond John's capabilities to do this alone. His most profound contribution to Christianity, therefore, was not in his development of baptism but in "nominat[ing] a successor who was far greater than himself." It was then Christ's duty to take up the work of John the Baptist, to develop it, complete it, and make it permanent.[9]

After situating Jesus within this renewed tradition of prophecy, Seeley next turned to a discussion of Christ's temptation, a key moment in Christ's life that apparently set the stage for all that would follow. Seeley argued that the temptation had long been misunderstood as a turning point in Christ's life when he was tempted by the Devil to use his powers to relieve his pain. What Seeley stressed, however, was that Christ's temptation was best understood as "the excitement of his mind which was caused by the nascent consciousness of supernatural power."[10] His period in the wilderness, therefore, was caused by his need to understand these new powers and to consider how he would go about using them, if at all. However, Seeley was insistent that Christ's temptation had nothing to do with the possible vanishing of morality under such an awesome power. Rather, what motivated Jesus was the need to establish "a set of newer and stricter obligations" that were necessitated by the possession of these new powers. Most pertinent is the fact that, after much struggle, Christ decided that he could not use this new power as a form of coercion, even if doing so could lead to the establishment of the Christian commonwealth that he envisioned. For Seeley, it was this ultimately moral decision that separated Christ from his messianic predecessors, such as David or Abraham, who chose a very different path in establishing their kingdoms. Jesus would not crush his enemies with his supernatural might but would rather rely on consent and moral persuasion.[11] It was from this moment on, according to Seeley, that Christ had a single plan and sought to execute it following the moral framework he had developed in the wilderness.

That plan was nothing less than the establishment of a kingdom of God. This idea would have, of course, been familiar to Christ's contemporaries. "Every Jew looked back to the time when Jehovah was regarded as the King of Israel," argued Seeley.[12] In proclaiming that "the kingdom of God is at hand," Christ was therefore appealing to a thousand-year-old tradition that included the Jewish statesman as prophet.[13] But even though Christ was appealing to a theocratic tradition of David and Solomon, the world had clearly changed. The "conditions of the age" were now very different, argued Seeley, and Christ recognized that he had to establish a kingdom under principles that were suited to those new conditions. "He saw that he must lead a life altogether different from that of David, that the pictures drawn by the prophets of an ideal Jewish king were coloured by the manners of the times in which they had lived; that those pictures bore indeed a certain resemblance to the truth, but that the work before him was far more complicated and more delicate than the wisest prophet had suspected." He therefore "revived the theocracy and the monarchy, but in a form not only unlike the system of David but utterly new and unprecedented."[14] Jesus had recognized that the times had changed and that the king would therefore need to be unlike any other.

So even though Christ appropriated the position of an ancient Jewish king, he declined many of the king's expected duties, such as commanding armies or presiding over law courts. Christ concerned himself only with "higher" duties such as would imply the "control over the wills of men."[15] Christ's contemporaries were rightly perplexed by this. And not only did Christ decline to engage in some of the functions of the Davidic monarchy, he also perhaps more importantly claimed for himself a position that the Davidic kings did not. As Seeley put it, "the kings of the house of David were representatives of the Invisible King in certain matters only. The greatest works which can be done for a nation by its shepherd were quite beyond their scope and province."[16] But it was precisely the "greatest works" that Christ claimed he could accomplish as part of a truly "heroic royalty."[17] Meanwhile Christ declined to control men the way the earthly kings did in the past, through coercion and by mobilizing armies.

While it is true that in proclaiming higher works for his rulership he therefore declined the role of civil judge, Seeley clarified that Jesus did in fact take "the diviner judgments, into his own hand."[18] Just prior to Christ's arrival, Jews began to contemplate the immortality of the soul. And as the later prophets postulated notions of immortality, Christ further developed that notion "into a glorious confidence." According to Seeley, "[t]his extension of the term of human life had a prodigious effect upon morality."[19] As a result, the civil courts became much less important than did the kind of eternal judgments Christ made on behalf of the invisible king, judgments that could lift souls to heaven or doom them to an eternity in hell.

It was at this point in the narrative that Seeley felt that, given that he had now fleshed out Christ's plan, or what he often referred to as "Christ's call," namely that Christ himself claimed to be the "Founder," "Legislator," and "Judge" of a new kingdom of God,[20] it was necessary to discuss the evidential basis for these claims as well as the claims to follow. This forced Seeley to engage in a rather perilous discussion of the Gospels, which he primarily relied upon to provide the essential facts of Christ's life. This was perilous, of course, because at the time the New Testament had been subjected to critical analysis by some of the

Essays and Reviews authors as well as other contributors to the historical Jesus genre, most notably Renan and Strauss. While many of these critiques did not go as far as Strauss's in determining that the sources could only be understood as presenting the mythological world view of the ancients, it is clear that at the very least the general trustworthiness of the Gospels as historical sources was being questioned. What Seeley therefore sought to do was to present his own method for preserving the integrity of the Gospels not as mere conduits into the mythology of the ancient world but as reliable historical sources capable of providing very real insights into what actually happened.

Indeed, Seeley began his methodological discussion of the Gospels by claiming that his own interpretation did not follow that of Strauss or the more extreme proponents of biblical criticism. He said that *Ecce Homo* "aims to show that the Christ of the Gospels is not mythical." Seeley was adamant "the character these biographies portray is in all its large features strikingly consistent, and at the same time so peculiar as to be altogether beyond the reach of invention both by individual genius and still more by what is called the 'consciousness of the age.'"[21] It was in grasping the consistency of the presentation of Christ through a comparison of the Gospels that Seeley was convinced that the character presented there must be true. Seeley argued, therefore, that he readily accepted those facts that were shared between the Gospels. He did, however, suggest that, given the differences between the first three Gospels and the fourth, he found it impossible to draw upon the fourth Gospel on its own.

Seeley's principle of accepting the facts that were shared between the Gospels put him in the awkward position of commenting on the issue of Christ's miracles, which are discussed there. Christ's miracles and what they have been interpreted as ultimately meaning was precisely the kind of discussion Seeley wanted to avoid by focusing instead on Christ's character. To get around this problem, Seeley argued that he would ignore, for the moment at least, "whether miracles were actually wrought," but would instead say that it was possible to establish "a fact which is fully capable of being established by ordinary evidence, and which is actually established by evidence as ample as any historical fact whatever – the fact, namely, that Christ *professed* to work miracles." He went on to say that "We may go further, and assert with confidence that Christ was believed by his followers really to work miracles, and that it was mainly on this account that they conceded to him the preeminent dignity and authority which he claimed."[22]

Clearly, for Seeley, ignoring the issue of Christ as a worker of miracles was not possible, given the method he decided to pursue in accepting that which was common between the Gospels. Moreover, he well knew from his father's work that many Christians believed strongly that the very basis of Christianity itself rested on the supernatural facts of the New Testament. But Seeley was not quite prepared to say that because miracles were described in the Gospels, they therefore physically happened. That was, according to Seeley's argument, beside the point. The fact remained that the authors of the Gospels believed they had witnessed miracles. So for Seeley's purpose, "which is to investigate the plan which Christ formed and the way in which he executed it, it matters nothing whether the miracles were real or imaginary; in either case, being believed to be real, they had the same effect. Provisionally therefore we may speak of them as real."[23]

Despite the convoluted nature of this methodological point, it speaks quite centrally to what Seeley was ultimately trying to achieve with his study of Christ. He wanted to show that the actual physical existence of certain supernatural occurrences was largely irrelevant to what really brought men and women together in worshipping Christ. And in this regard what is most necessary to understand is all of Christ's attributes, words, and actions. "It was neither for his miracles nor for the beauty of his doctrine that Christ was worshipped," Seeley argued. "Nor was it for his winning personal character, nor for the persecutions he endured, nor for his martyrdom. It was for the inimitable unity which all the things made when taken together."[24] For Seeley, it was the person of Christ himself that men and women decided to worship, not Christ as the worker of miracles, or Christ as a martyr. "Witnessing his sufferings, and convinced by the miracles they saw him work that they were voluntarily endured, men's hearts were touched, and pity for weakness blending strangely with wondering admiration of unlimited power, an agitating of gratitude, sympathy, and astonishment, such as nothing else could ever excite, sprang up in them, and when, turning from his deeds to his words, they found this very self-denial which had guided his own life prescribed as the principle which should guide theirs, gratitude broke forth in joyful obedience, self-denial produced self-denial, and the Law and Law-Giver together were enshrined in their inmost hearts for inseparable veneration."[25]

Seeley admitted that it was not an easy task to understand just why Christ's contemporaries were drawn to him and chose to follow his call. In his own time, that is the nineteenth century, Seeley claimed that

Christians were often defined by how they positioned themselves in relation to a variety of theological doctrines such as the Atonement or the Resurrection. But what modern theologians failed to grasp, Seeley argued, was that early Christians would have been utterly baffled by these concepts. The creed of the first Christians was of an elementary character and it is therefore difficult to comprehend from the perspective of the present, which is shaped by a more elaborate theological understanding of Christianity.[26] Here Seeley stressed the importance of his historical training in enabling such a nuanced understanding. "Only a well-trained historical imagination," Seeley argued, "active and yet calm, is competent so to revive the circumstances of place and time in which the words were delivered as to draw from them, at a distance of eighteen hundred years, a meaning tolerably like that which they conveyed to those who heard them."[27] And from the perspective of a "well-trained historical imagination," it was important to understand that by the nineteenth century, Christianity had gone through several transformations as it expanded and was reinterpreted in different historical contexts. As it was adapted to these new contexts, it needed to rely on new methods of justification. For Seeley, these more modern methods of justification were skewing a properly historical understanding of what was actually required of early Christian believers. It was simply faith in himself that Christ demanded, not a belief in particular theological doctrines.[28] So what was Christ's character, which Seeley deemed so important to early Christianity?

Seeley answers this question in the second part of the book, which is ostensibly about "Christ's legislation." The use of this terminology is curious, given that Seeley argued that Christ did not establish any explicit laws in the same way as did his Davidic predecessors. But what soon becomes clear is that Christ's "legislation" was nothing other than his character. His deeds became precedents to follow. The Sermon on the Mount is central in this context for Seeley, as it illustrates Christ's character as one where "thoughts, words, and deeds are to be of a piece."[29] Christ showed that we must love our neighbour not because we will be punished if we do not or because we will not gain entrance to heaven but rather because love in itself is virtuous. "By defining virtue to consist in *love*," argued Seeley, Christ "brings into prominence its unselfish character, and by denouncing at the same time with vehemence all insincerity and hypocrisy, he sufficiently shows with what horror he would have regarded any interested beneficence or calculating philanthropy which may usurp the name of love."[30] Hence Christ's kingdom was "a brotherhood founded in devotion and self-sacrifice."

In promoting a "disinterested devotion" to one's neighbour, Seeley argued, Christ was appealing to something approximating patriotism. Patriotism was a necessary feature of political societies, according to Seeley, and was really an extension of the bonds that united small families at the origins of human social relations. As family units expanded and then formed alliances with other families, much larger political entities, such as nations, began to appear. Seeley claimed that this historical development of patriotism and politics was precisely how the Jewish nation was forged. The language of bloodlines and familial bonds remained central to its members: "[t]heir common descent from Abraham was always present to their minds, and the tie which bound them together."[31] What Christ did was to make this familial metaphor universal, by stating that all men are brothers, descended not from Abraham but from God. "By substituting the Father in Heaven for father Abraham, Christ made morality universal." Seeley admitted that this notion may not have been entirely original, "But to work it into the hearts and consciences of men required a much higher and rarer power, the power of a ruler, not of a philosopher."[32] This, then, is the first law of the Kingdom of God, "that all men however divided from each other by blood or language, have certain mutual duties arising out of their common relation to God."[33] Under this scheme there are no longer any bonds that take precedence over the bond that all humans share under God. "Christ declares all men alike to be the sons of God," argued Seeley, "and the least of mankind he adopts as a brother." Christ therefore "makes all mankind equal to this extent, that the interests and the happiness of all members of the race are declared to be of equal importance."[34] "To love one's neighbour as oneself was, he said, the first and greatest *law*."[35]

It is in the discussion of this first "law" that we come to the main argument of *Ecce Homo*. It is one thing, Seeley argues, to say that men should love their neighbours as they love themselves, but it is quite another to convince men actually to believe it. As well as creating this new morality, Christ had to change man's state of mind, to convince him to love his neighbour. And as his great temptation in the wilderness made clear, this could not be done by force, or by the use of his supernatural powers, but rather by his own character. "[W]hen the precept of love has been given," argued Seeley, "an image must be set before the eyes of those who are called upon to obey it, an ideal or type of man which may be noble and amiable enough to raise the whole race and make the meanest member of it sacred with reflected glory." It was

Christ, of course, who did this. He gave in his words and deeds a living example of the ideal of humanity.[36] It was in this way that Christ kindled, according to Seeley, "an enthusiasm for humanity."[37] By showing that it was possible to not only love one's neighbour but even to love one's enemy, Christ "raise[d] the love of man as man to enthusiasm." It was this unreserved love that Christ had for the ideal of man that was infectious. This, for Seeley, was the "very kernel of the Christian moral scheme."[38]

It was the very kernel of Christianity for Seeley because it followed from Christ's enthusiasm of humanity, from his love of the ideal of man in *every* man, that there need not be any negatively defined laws determining how people act towards one another. Every member of the Christian commonwealth, who feels this same enthusiasm, "is a law-giver to himself." The "divine inspiration" that is felt from this enthusiasm, that is "by contemplation of Christ's character," necessarily "dictates ... the right course of action." So it is not Christian *law* that is paramount for Seeley in attempting to understand Christ's society, but the Christian *character*, namely "the new views, feelings, and habits produced in the Christian by his guiding enthusiasm."[39] These new feelings resulted in ultimately a "moral reformation" that Seeley summarized as transforming society from being morally restrained to being morally motivated. Whereas the morality of the Old Testament was based on prohibitions (e.g., "thou shalt not"), the morality of early Christianity was based on "positive" affirmations (e.g., "thou shalt").[40] In helping to establish this new morality, Christ's followers were not therefore merely doing what they were told. They were following Christ's example for how to live and how to treat their fellow men. Christ therefore "set the first and greatest example of a life wholly governed and guided by the passion of humanity."[41]

That Christ ignited an "enthusiasm for humanity" was the central claim of *Ecce Homo*. While this claim certainly helped establish the originality of the book, it also goes some way to expose its author's own intellectual and historical context. To say that Christ kindled an "enthusiasm," of course, is not terribly original, given that the concept has deep Christian connotations. But at the time it was a label often pinned on evangelicals (sometimes by their enemies) as a way to describe their particularly passionate, even irrational faith. So Seeley was appropriating a term from his own religious past while pairing it with a term of a much more modern vintage, namely with "humanity." This was a bold move indeed. That is because the term "humanity" connected Seeley's conception of Christ very clearly with an aspect of positivist philosophy that was gaining prominence precisely when *Ecce Homo* appeared. Seeley was ultimately trying to indicate that this seemingly modern and secular philosophy of humanity was not just complementary to Christianity but actually present at its birth, no better symbolized than in the figure of Christ himself. By suggesting that Christ above all had an "enthusiasm for humanity," Seeley was ultimately arguing that Jesus wanted to establish a thoroughly integrated society based on Christian principles that would bring together a society's religious and political lives. It would be a society based on a profound "love of man," one that is ethically moral, sympathetic, and, more than anything, "humane."[42]

Moreover, and excluding the specifically religious connotations, positivism above all stood for a progressive understanding of the development of human society. And for Seeley, Christ's enthusiasm of humanity engendered a revolution in the realm of morality that was clearly a progressive development from what preceded it. This was not in itself a radical interpretation, but it also implied that Christianity was still developing and that part of its further development depended on grasping the kernel of what it was that Christ really achieved. This is particularly clear in Seeley's discussions of the subsequent "laws" that naturally developed from Christ's enthusiasm of humanity.

One of the first such laws to develop out of Christ's love of man was what Seeley called the "Law of Philanthropy." In Christ's day, as Seeley argued, this law symbolized the charitable obligations Christians felt towards their fellow man, such as aiding the sick or giving to the poor. Seeley stressed that there were acts of charity before Christ. But these were much more limited, typically offered to family members or to fellow members of a closed society, not to humanity as a whole. So the philanthropy that developed in Christ's day was a great advancement. And yet it was still quite limited in its scope, owing to the "conditions of that age." Because it followed so closely from an era of barbarism, the philanthropy of Christ's age "was faint and feeble in its enterprises, the half-despairing attempt of a generation which had more love than hope."[43] Such acts of charity, according to Seeley, would be utterly insufficient in later historical periods, and therefore would not live up to the spirit that led to the law in the first place. It was Seeley's point that the law of philanthropy is not static; that it changes as society changes; and that it therefore should not be limited by the ancient social knowledge that accompanied its birth. We are no longer limited, Seeley argued, by the knowledge of the ancient world, and have "at our disposal a vast treasure of science, from which we may discover what physical well-being is and on what conditions it depends. In these circumstances the Gospel precepts of philanthropy become utterly insufficient." If, therefore, Christ initially called on "his first followers to heal the sick and give alms," Seeley claimed that Christ now

> commands the Christians of this age … to investigate the causes of all physical evil, to master the science of health, to consider the question of education with a view to health, the question of labour with a view to health, the question of trade with a view to health; and while all these investigations are made, with free expense of energy and times and means, to work out the rearrangement of human life in accordance with the results they give. Thus ought the Enthusiasm of Humanity to work in these days and thus, plainly enough, it does work.[44]

Christ would therefore be a positivist humanitarian in the nineteenth-century sense, seeking to apply the findings of modern science to the betterment of society as a whole.

It is at this point in *Ecce Homo* that it becomes abundantly clear why lengthy discussions of specific passages in the New Testament or of theological doctrines have been largely absent. This is because, for Seeley, Christianity has never been a static religion. It was developed out of ancient Jewish practices and beliefs into a society in transition. Its apparent doctrines, therefore, are nothing but ephemeral vestiges, the necessary result of past historical circumstances. "The New Testament is not the Christian law," Seeley argued; "the precepts of Apostles, the special commands of Christ, are not the Christian law. To make them such is to throw the Church back into that legal system from which Christ would have set it free." But what is eternal about Christianity, according to Seeley,

> is the spirit of Christ, that Enthusiasm of Humanity which he declared to be the source from which all right action flows. What it dictates, and that alone, is law for the Christian. And if the progress of science and civilization has put into our hands the means of benefiting our kind more and more comprehensively than the first Christians could hope to do … we are not to enquire whether the New Testament commands us to use the means, but whether the spirit of humanity commands it.[45]

If the spirit of enthusiasm commanded men in the ancient world to ease the suffering of the sick, it commands men of the modern world to prevent the sickness in the first place. It is, in other words, the spirit of humanity that modern Christians should be following, the spirit developed from Christ's character, and not any particular commandment that may be found in the New Testament.

Seeley engaged in similar discussions of how the laws of edification, mercy, resentment, and forgiveness were born out of Christ's enthusiasm of humanity, how they represented an advancement from what came before, and how they developed in the context of modern civilization. The doctrine of forgiveness is perhaps the most important of these, in part because, as Seeley remarks, it is "the most *distinctive* feature in the system," and it is often understood as "*characterizing* Christian morality more than any other doctrine." Seeley argued that Christ's concept of forgiveness marked a profound distinction between the ancient and the modern world. It was not that the ancients did not know of forgiveness or consider it a virtue, but they also valued the necessity of revenge. Christ, by contrast, preached what Seeley referred to as *unlimited* forgiveness, that is the kind of forgiveness offered to one's enemies even as a result of personal injury. This is a truly new virtue, according to Seeley, and one that signifies the merging of Christianity with civilization itself, no better illustrated than in the fact that in modern societies revenge is taken out of the hands of the injured party and placed in the hands of juries and judges.[46]

Seeley's Jesus, therefore, was what we might call the moral Jesus, or the Victorian Jesus, a man whose love of humanity was meant to provide an example for all others to live by while establishing a thoroughly integrated society whereby everyone could pursue their happiness without fear of their fellow man. For Seeley, this view of Jesus had been profoundly obscured and had unfortunately lost much of its "attractive power." "The prevalent feeling towards him now among religious men is an awful fear of his supernatural greatness, and a disposition to obey the commands arising partly from dread of future punishment and hope of reward, and partly from a nobler feeling of loyalty, which, however, is inspired rather by his office than his person."[47] It was time, therefore, to remind Christians about Jesus the person as opposed to Jesus the supernatural creator (and destroyer) of worlds. And, indeed, bring him to life within a nineteenth-century political and social context.

This was certainly a "positive" theology, which many were calling for in the wake of the crises surrounding the publication of *Essays and Reviews* and Colenso's studies of the Old Testament. And, in many ways, Seeley was following the writing practices of his father by attempting to show that fairly orthodox interpretations of Christian history need not be in conflict with contemporary scientific advances. Where father and son differed, however, was in the fact that they wrote for different kinds of Christian audiences. While the father spoke directly to evangelical believers in order to reassure their already-existing faith, the son spoke specifically to those convinced by the new rationalist methods, adherents of biblical criticism, positivism, and even evolutionists, who were growing sceptical of the established Church or who were already well on their way to some form of unbelief. Seeley no doubt knew that his father would have likely disagreed with his study of Christ's life. It certainly did not conform to the biblical literalism favoured by the older man. At the same time, Seeley sought to avoid the really contentious theological issues that divided Christians in order to focus on what might unite them. What mattered then, for Seeley, was not if Christians could agree on whether Christ was resurrected or if his death was a symbol of his atonement for the sins of humanity but rather if they could agree about the adaptable set of morals that almost naturally developed from Christ's "enthusiasm of humanity." As Seeley argued in the closing pages, "The story of his life will always remain the one record in which the moral perfection of man stands revealed in its root and unity, the hidden spring made palpably manifest by which the whole machine is moved."[48] For Seeley, it was the story of Christ's moral character that truly mattered, and he hoped that focusing on this most important aspect of Christ's life would help engender a discussion about establishing a reformed Christian Church, one modelled on these moral principles, which would be thoroughly updated thanks to the findings of modern science.

How Seeley was to go about reaching his chosen audience with his important message was another issue entirely. He, of course, needed a publisher. In other circumstances, his father's publishing house might have been a good choice. But the message of *Ecce Homo*, along with its anticipated readership, would not be suited to Seeley and Burnside. He did, however, know of one publishing firm that was run by former employees of his father. And it published works for precisely the educated, liberal, and secularizing audience Seeley wanted to reach. In terms of authorship, Seeley also followed in his father's footsteps by having his intervention into the religious debates of the day published anonymously. Also like his father, he sought to utilize his anonymous authorship as a persona that could be trusted to write honestly on such a highly contentious subject.

However, by 1865 the practice of anonymity was no longer as widely accepted a norm of the publishing world as it was in the first half of the nineteenth century. Anonymous publishing began to be questioned in a variety of genres and mediums, as readers and critics began to wonder about the purposes of hiding one's identity in an age of liberal openness. Whereas Seeley's father likely received little scrutiny as "the author of *Essays on the Church*," the same could not be said for the author of *Ecce Homo*, whose integrity and true authorial intentions were often openly questioned as readers tried to grasp what was clearly a new but often ambiguous, challenging, and, at times, radical message to receive. It is to these interrelated issues that we now turn.

Chapter Five

A Dangerous Book

All the unconscious guiding which a name, even if hitherto unknown, gives to opinion is wanting.

– "Ecce Homo," *Guardian*, 7 February 1866

"This is a dangerous book to review," announced the opening line of a review that was published in the weekly *Guardian* on 7 February 1866. The book in question was *Ecce Homo: A Survey in the Life and Work of Jesus Christ*. It was dangerous in part because of the subject matter. As we have seen, histories of Jesus were intensely controversial at the time. And the religious establishment was still dealing with the legal ramifications surrounding the publishing of *Essays and Reviews* and Bishop Colenso's studies of the Old Testament. Seeley's attempt to steer clear of much of this controversy, by simply avoiding any discussion of theological matter at all, backfired in many ways, as the theological ambiguity left many readers uncomfortable and confused about the author's purpose. And this was compounded by the fact that the book was published anonymously, thereby depriving readers of the ability to deduce some sort of meaning based on the author's identity. This was really what the reviewer of the *Guardian* was getting at when he said that the book was a dangerous one to review: As he went on to say, "All the unconscious guiding which a name, even if hitherto unknown, gives to opinion is wanting."[1] This did not stop readers from trying to discover the identity of the author in order to find the hidden meanings of the text, however. To the contrary. *Ecce Homo* became a literary sensation in 1866 in large part because the public was invited to speculate on the author's identity and in doing so to project particular meanings, fears, and desires onto a book that was in theory supposed to transcend traditional party allegiances.

In this regard, the *Guardian* review was a masterful behind-the-scenes effort by both Seeley and his publisher, Alexander Macmillan. Through back channels, they saw to it that a *Guardian* review would be written that would comment on the perceived problems of *Ecce Homo* while turning them into strengths. The *Guardian* review recognized that readers would be concerned that the author did not identify himself. At the same time, the reviewer stressed that the anonymous author could be entrusted to take the reader on this journey of religious discovery. As we will see, the *Guardian* review was an enormous success for author and publisher: it changed the conversation about the book, which was previously focusing on what many found to be its problematic methodology. Moreover, the *Guardian* was an important orthodox Anglican voice that acted to legitimate the publishing of such a dangerous book. Just how the reviewer of the *Guardian* was convinced to provide such an endorsement is the subject of this chapter, the purpose of which is to demonstrate how Seeley and his publisher worked together to shape the early reception of the book.

The *Guardian* review was published about three months after *Ecce Homo* first appeared. "This day is published," announced a Macmillan and Co. advertisement in the weekly *Reader* on 25 November 1865, "ECCE HOMO! A SURVEY in the LIFE and WORK of JESUS CHRIST." Similar advertisements were taken out in the *Spectator*, the *Examiner*, and the *Athenaeum*, in an attempt to alert the educated general reader of the existence of the book, which began with an initially modest print run of fifteen hundred copies.[2] Seeley must have been exceedingly happy with his publisher, who was giving his newly published *Ecce Homo* top billing in the many advertisements for new books. Indeed, at this early stage, publisher and author seem to have been on the same page, as it were, when it came to the marketing and selling of Seeley's anonymous study.

It is certainly tempting to suggest that Seeley chose for his study of Jesus a publishing firm that reflected his Broad Church outlook. However, he had a longstanding personal relationship with the house of Macmillan that at least initially had nothing to do with their similar religious outlooks. And even before Seeley had any relations with Macmillan and Co., the Macmillans were familiar with Seeley's father. Indeed, well in advance of the establishment of Macmillan and Co. in 1843, its founders, Daniel and Alexander Macmillan, had worked for Seeley and Burnside in the late 1830s and early 1840s. This was Daniel's second publishing firm, but Alexander was just twenty-one when he was hired by Seeley and Burnside and learned the trade under R.B. Seeley's employ.[3]

It was while working at Seeley and Burnside that Daniel and Alexander began to recognize that they could go out on their own and establish their own publishing house. One might have thought that their upbringing on a farm in Scotland, with no formal education, would have worked against the Macmillan brothers. However, they were hard workers with a great deal of ambition, and they read extremely widely to make up for their educational deficiencies. After saving enough funds and securing a small loan, they were able to open up in London their own small publishing house in February 1843. Later in the year they published their first two books: A.R. Craig's *The Philosophy of Training* and William Haig Miller's *The Three Questions: What Am I? Whence Came I? And Whither Do I Go?* Miller had previously written for the evangelical Religious Tract Society, which suggests that the Macmillans were able to tap into Seeley and Burnside's network of authors. Moreover, as Rosemary VanArsdel has argued, these two books established long-standing "trends in Macmillan publishing philosophy which would influence their lists far into the future: sincere interest in education and profound religious belief."[4]

The Macmillans soon relocated to Cambridge, which proved extremely beneficial. There they came into contact with many university students, dons, and intellectuals. Moreover, their publishing list expanded exponentially. Not only did they begin publishing works specifically for Cambridge students but they also began publishing the works of several high-profile Cambridge clergymen, most notably those associated with the Broad Church movement. F.D. Maurice became an early supporter of the Macmillans, publishing several books and pamphlets on many of the religious questions of the day.[5] For instance, the debate that Seeley and his father had about Maurice's views of atonement, discussed in chapter 3, was in large part about Maurice's *Theological Essays*, a controversial book published by Macmillan and Co. in 1853. They published several of Charles Kingsley's works, notably *Westward Ho!* (1855), which proved to be remarkably profitable. They also published Thomas Hughes's anonymous *Tom Brown's School Days* (1857) and became a significant publisher of science works as well.

The wide range of interests pursued by the publishing firm was also nicely signified by the publishing emblem, which was commissioned in 1861, and adorns the cover of this book. The central image is of a cross, with a double M in the middle that represents the two Macmillan brothers. The cross also includes four symbols: three stars at the top, a bee on the right, a butterfly on the left, and three acorns at the bottom. In commissioning the emblem, Alexander Macmillan chose the symbols quite specifically. "The stars," Macmillan explained, were "for heavenly glory and light; the acorns for earthly growth and strength; the bee for useful industry; the butterfly for beauty, pure and aimless."[6] This emblem would appear in Macmillan books for most of the nineteenth century and was carved over the entrance of the firm's London shop.[7]

It was while the firm was still primarily in Cambridge, however, in the 1850s, that the Macmillan shop became well known as a vibrant space where debates and conversations on a wide range of topics took place among Macmillan authors, public intellectuals, and friends. It was during this period that John Seeley became familiar with the Macmillans. Writing to Alexander Macmillan in 1882, John reminisced that his father had taken him to visit the Macmillans in October 1852 when he had just entered Christ's College, which demonstrates that the Macmillans maintained a relationship with the elder Seeley well after they left his publishing business. John remembered in particular the engaging discussion that took place during their visit, writing that "I was too shy to contribute anything myself."[8] So began Seeley's longtime friendship with Alexander Macmillan. It was at this point, too, that Seeley entered into a circle of friends who all ended up sympathizing with the Broad Church movement, such as Edwin A. Abbott, J. Llewelyn Davies, and John Venn.[9] That Seeley and Alexander became close while Seeley was in Cambridge there is no doubt. When Daniel finally succumbed to a long and drawn out struggle with tuberculosis in 1857, Seeley remembered Alexander finding him playing cricket in order to let him know that Daniel had literally just died.[10]

After Daniel's death, Alexander took over the running of the publishing business and soon after opened a London branch in Covent Garden. The informal intellectual gatherings that had taken place in Cambridge now became formalized in London and would occur on Thursdays when Macmillan made his weekly trip there. These gatherings became known as "Tobacco Parliaments" and included an impressive and diverse array of London's leading literati. "Men of such diverse tastes and talents as Alfred Tennyson, Herbert Spencer, T.H. Huxley, Francis Turner Palgrave, Coventry Patmore, F.D. Maurice, Charles Kingsley, and Thomas Hughes gathered for a light supper, followed by tobacco and spirits, to engage the topics of the day in this leisurely setting."[11] There is little doubt that John Seeley would have also been a regular at these gatherings. For he had made the move to London as well. In 1859 he became the chief classical assistant master at the City of London School. And in 1863 he became professor of Latin at University College, London. Sending his manuscript of *Ecce Homo* to Macmillan was, therefore, an obvious choice. It is likely that Seeley spoke to Macmillan about his project well in advance of it being considered for publication. It must be understood, therefore, that when *Ecce Homo* was published, Macmillan and Seeley were more than just two individuals engaged in the business of publishing a particular book. They were close friends with a mutual history that went back well over a decade. At least early on, the fact of their close relationship seemed to work well in regard to the shaping of the book's reception.

Indeed, as *Ecce Homo* hit the bookshops in November of 1865 with no authorial presence on the cover page, author and publisher had to maintain fairly close relations. This was in part because all correspondence about the book went through the publisher, meaning that Macmillan relayed any and all messages that were directed to the author of *Ecce Homo*. Often this simply involved forwarding such responses to the book as enclosures in letters from Macmillan. For instance, on 13 December Macmillan received the first of many letters that he would forward on to Seeley that would praise the "remarkable" *Ecce Homo*. "I have been reading E.H. with intense interest," the principal of Cheshunt College, Dr Reynolds, wrote to Macmillan, "and feel half dazzled by it. I have seldom seen a greater condensation of thought and suggestion, nor a more startling presentation of harsh truth."[12] But, perhaps more importantly, Seeley and Macmillan had to stay in close contact with regard to shaping the reception of the book. Responding to criticism is slightly more complicated when the author's identity is unknown. And as we will see, a particularly problematic issue arose soon after *Ecce Homo* was published, having to do with the fact that there was no authorial persona attached to the volume.

Initially, however, there had been some anxiety on Seeley's part that *Ecce Homo* was going to be ignored entirely. It did not help matters that no high-profile reviews of the book appeared for nearly two months. "Richmond [Seeley's brother] told me on Saturday that you were disappointed in not getting more reviews," Macmillan wrote to Seeley. "You may keep your mind easy. You staggered the critics and they have not recovered their health yet. You will hear their sweet voices by and by."[13]

And indeed Seeley would, though "sweet" would not have been the adjective he would have used to describe them.

Frustrating Seeley further was the fact that the first few reviews that trickled out in December, January, and early February, while generally mixed in their overall assessment, were misleading and critical of some of the basic premises underpinning the investigation. This problem was particularly pronounced in the first review of *Ecce Homo*, published in the widely circulated weekly the *Spectator*, though the review appears at first glance entirely positive. Macmillan admitted that the review was for the most part "excellent," given the way in which it described "the power and originality of the book." It was penned by the *Spectator*'s editor, R.H. Hutton, who had a longstanding relationship with Macmillan. Hutton found the book "full of striking thought and delicate perception, a book that has realised with wonderful vigour and freshness – with far more power than Neander, and far more both of power and truth than Renan and Strauss – the historical magnitude of Christ's work, and which here and there gives us readings of the finest kind of the probable motive of his individual words and actions."[14] But this generally positive tone was undermined by a single rather destructive criticism that would haunt the entire reception of the book.

Indeed, "the only great fault" Hutton articulated was so central to the very core of the book that both Macmillan and Seeley felt that the review would be quite damaging. "The attempt to delineate from within the life of Christ, without making any fundamental theological assumption as to its nature and the reality of his revelations," read the first line of the review, "is almost like the attempt to paint a picture without making any assumption as to the quarter whence the light comes, and consequently whither the shadows fall."[15] The author's "defect of method" was an unfortunate one, argued Hutton, because it "veil[s] in great measure the subject of his inquiry – Christ himself – from himself and his readers."[16] This criticism indicated to Macmillan that Hutton simply "missed its [*Ecce Homo*'s] real point. The great fault which he finds is surely the very essence of your idea in its mode," Macmillan wrote to Seeley. This was concerning because Macmillan "feared that the general public would fail to see this idea," and he had "quite looked to Hutton for bringing out ... the value and significance of your mode of treating their great subject." This was why Macmillan had given the lead to Hutton at the expense of other reviewers, though in hindsight this was clearly a mistake. Macmillan was certainly justified in worrying about the afterlife of Hutton's criticism, and much of his early strategy in promoting the book was devoted to challenging any notion of there being a central methodological defect at work in *Ecce Homo*. For now, Macmillan decided to write to Hutton in order to convince him of his mistake. He also wondered if Seeley would consider writing to Hutton himself.[17]

Seeley apparently did, because a formal "letter to the Editor of the 'Spectator'" was published in the *Spectator*'s pages the following week – signed by "The Author of 'Ecce Homo'" (see figure 5.1). In his critique of what was an admittedly favourable review, Seeley asked the *Spectator* to "reconsider part of it." The review "misconceived my plans, and I am seriously afraid that the misconception may spread to those whom your paper influences." Seeley explained that what the review seemed to misunderstand was that *Ecce Homo* was by no means the result of a lifelong inquiry into the early history of Christianity. *Ecce Homo* was not a patient analysis "written after the investigation was completed, but the *investigation itself*." *Ecce Homo* was therefore a study-in-progress, which was why "I have postponed altogether the hardest questions connected with Christ, as questions which cannot properly be discussed until a considerable quantity of evidence has been gathered about his character and views."

This letter highlights the main strategy that Seeley and Macmillan would employ in dealing with challenges to the methodology, that the theological issues were not ignored but simply put aside and postponed for a future study that would necessarily build on the historical foundation provided by this initial investigation. "But pray do not suppose that postponing questions is only another name for evading them," Seeley wrote in his letter. "I think I have gained much by this postponement."[18] The *Spectator*'s rejoinder to the letter admitted that "Our criticism was perhaps written under some slight misunderstanding as to what the author really intended his book to be," but was adamant that the central criticism of the review was valid, that divorcing Christ's life from any theological considerations led the author to misconceive the "leading aim of Christ," which was not, as *Ecce Homo* claimed, an "enthusiasm of humanity." Hutton also said as much in his personal letter of reply.[19]

Initially at least, Seeley's letter to the *Spectator* did not have the desired effect. This is in part because, before it could be published, other reviews weighed in. The dissenting *Patriot*'s review, for instance, not only reflected the opinion of the *Spectator*, it showed quite clearly how such criticism could be utilized to denounce the book *in toto*.

had his knife in it." He took no notice of the remark, which drew all eyes upon him and upon the lady, but by and bye she stretched out her hand, and took from the plate some chipped dried beef, which stood between her and her victim. This was well enough, of course; but he turned at once, and calling a waiter said, only as if he were asking for more tea, "Take away that dried beef, this lady has had her fingers in it." In this encounter, such as it was, he was thought to have had the best of it, and she did not forgive or forget. So a few days afterwards (I should have mentioned that there was the slightest possible acquaintance between them) they being at dinner, she, conspicuous in the full dress she had adopted since her tour to Europe, and which was so very "full" that it would have attracted attention under any circumstances, took one from a dish of fresh figs before her, and putting it on a plate, handed it to him with an expression of complaisance, but saying in a tone of unmistakeable significance, which could be heard all around her, "A fig for you, Sir." He accepted it graciously, and taking in his turn a leaf from the garniture of the dish, offered it to her, with "A fig-leaf for you, madam." She fled the table, and kept her room until her intended victim left the hotel. It was generally agreed that he had done what a gentleman would shrink from doing; but the provocation was such that he was held guiltless of offence, and applauded for his wit; and nobody, except perhaps a few of her slaveholding friends, pitied her. This is the last instance that I know of, of a lady's appearance in full dress at a public table. But I am told that within the last three or four years it is coming more in fashion among the "fastest" sets at the height of the season at Saratoga, and one or two others of our gayest watering-places.

A knowledge of affairs in this country may be conveyed to readers in England as well, perhaps, by the correction of erroneous statements made there with an air of authority as by independent accounts of what is going on here. It is for this reason chiefly, I might almost say for this reason only, that I have noticed a very few of the multitudinous misrepresentations of us and our affairs that I have encountered in British books and periodicals. I came across several at one sitting yesterday, some of which were fine examples of the degree of ignorance and the indifference to knowledge of the United States on the part of those who undertake to instruct you on that subject. The *Edinburgh Review* for October opens with an article upon what it styles "American Psychomancy," but which is nothing more nor less than that childish folly, spirit-rapping and its concomitants, with this paragraph:—

"About midway between New York and Albany, on the eastern bank of the Hudson river, stands the pleasant town of Poughkeepsie, containing a population of nearly twenty thousand souls. A quarter of a century ago the site was occupied by a few miserable cottages and farmsteads, and a solitary building for public worship. It now includes many handsome rectangular streets, sixteen churches, four banks, various large factories, an endowed collegiate school for boys, a corresponding academy for girls, and the Pantheon of Progress."

Now here is a distinct statement made about a physical fact, correct information as to which is easily accessible; and it is made too, not by an irresponsible penny-a-liner, or a man who writes a leading article at night which must be published on the morrow, but by a quarterly reviewer, who makes it the starting point of a grave discussion, and who should certainly have taken a little care not to assert what he did not know to be true. And what is the truth? It is that (as may be seen by *Morse's Gazetteer*, published in London, 1798), more than three-quarters of a century ago, in 1789, Poughkeepsie (the name is a corruption of Apo-keep-sing, meaning a pleasant harbour) was a village of 2,529 inhabitants, having then two churches, one of them Episcopalian (as the Church of England is called here), the existence of which is a sure indication in this country of the higher forms of social and intellectual culture. In 1820, forty-five years ago, this place, the site of which, according to the *Edinburgh* reviewer, was twenty-five years ago occupied by a few miserable cottages and farmsteads, had 5,726 inhabitants, 1,448 children in its schools, 9 flour mills, 30 fulling mills, 5 cotton and woollen factories, 5 churches, an academy, and a Lancastrian school. It was the county town, too, of the richest rural county in the State—Duchess—and of course contained the county buildings, which were substantial stone structures, and the county bank. In 1840, just the reviewer's quarter of a century ago, the place had 10,000 inhabitants, with more than a corresponding increase of prosperity and importance; and for this reason. The township and the country round are filled with the rich farms and the large county seats of men of hereditary wealth and culture, whose places are on such a scale and kept up in such a style as to delight and surprise (as I know) visitors from Europe who belong to those classes whose habits of life make them the most exacting critics in such matters. And in fact this "site of a few miserable cottages and farmsteads" was, "a quarter of a century ago, as it is now, the place where, if at all, might be found such elegant and highly cultivated do nothings as the Parapets in *Marian Rooke*. Now all this is of comparatively little moment or interest, except as data from which to infer the trustworthiness of a reviewer, the object of whose paper is to show that, in the words which he adopts from some crazy or crafty spiritualist, who lives, like Simon Magus, upon his shallow pretensions to sorcery, "half the members of Congress and the State Legislatures as well as half the scientific and literary men of America, are spiritualists." Putting his condemnation in the form of commiseration, he exclaims, "Well may the sober in the States stand aghast at such professions and practices as these, and despond for the future of their country."

A YANKEE.

"ECCE HOMO."

[TO THE EDITOR OF THE "SPECTATOR."]

SIR,—The perfect candour and cordiality of your review of *Ecce Homo* emboldens me to ask you to reconsider part of it. From your critical opinion I should not appeal, indeed it is so favourable that I should have no interest in appealing from it. But you have misconceived my plan, and I am seriously afraid that the misconception may spread to those whom your paper influences.

I endeavoured in my preface to describe the state of mind in which I undertook my task. I said that the character and objects of Christ were at that time altogether incomprehensible to me, and that I wished to try whether an independent investigation would relieve my perplexity. Perhaps I did not distinctly enough state that *Ecce Homo* is not a book written after the investigation was completed, but the *investigation itself*.

The *Life of Christ* is partly easy to understand and partly difficult. This being so, what would a man do who wished to study it methodically? Naturally he would take the easy part first. He would collect, arrange, and carefully consider all the facts which are simple, and until he had done this he would carefully avoid all those parts of his subject which are obscure, and which cannot be explained without making bold hypotheses. By this course he would limit the problem, and in the meanwhile arrive at a probable opinion concerning the veracity of the documents, and concerning the characteristics, both intellectual and moral, of the person whose high pretensions he wished to investigate.

This is what I have done. I have postponed altogether the hardest questions connected with Christ, as questions which cannot properly be discussed until a considerable quantity of evidence has been gathered about his character and views. If this evidence when collected had appeared to be altogether conflicting and inconsistent, I should have been saved the trouble of proceeding any further; I should have said Christ is a myth. If it had been consistent, and had disclosed to me a person of mean or ambitious aims, I should have said Christ is a deceiver. Again, if it had exhibited a person of weak understanding and strong impulsive sensibility, I should have said Christ is a bewildered enthusiast.

In all these cases you perceive my method would have saved me a great deal of trouble. As it is, I certainly feel bound to go on, though, as I say in my preface, my progress will necessarily be slow. I am much engaged, and have little time for theological study. But pray do not suppose that postponing questions is only another name for evading them. I think I have gained much by this postponement. I have now a very definite notion of Christ's character and that of his followers. I shall be able to judge how far he was likely to deceive himself or them. It is possible I may have put others, who can command more time than I, in a condition to take up the subject where for the present I leave it.

You say my picture suffers by my method. But *Ecce Homo* is not a picture; it is the very opposite of a picture; it is an analysis. It may be, you will answer that the title suggests a picture. This may perhaps be true, and if so it is no doubt a fault, but a fault in the title, not in the book. For titles are put to books, not books to titles.—I am, Sir, your obedient servant,

THE AUTHOR OF "ECCE HOMO."

[Our criticism was perhaps written under some slight misunderstanding as to what the author really intended his book to be. But we did not mean so much to complain that *Ecce Homo* is not a picture of Christ, as that the analysis, on which the author lays so much stress, of the leading aim of Christ—the intention

Figure 5.1 "To the Editor of the Spectator" signed by "The Author of 'Ecce Homo.'" *Spectator*, 30 December 1865, p. 1467. Courtesy of the *Spectator*.

The *Patriot* lamented that *Ecce Homo* "presents us with a view of the Saviour's teaching and work, from which some of its most precious truths and most distinctive features are all eliminated, and His character as a Redeemer, if not absolutely denied, quietly ignored." While the reviewer admitted that there were some remarks about Christ and the founding of his great church that were admirable, "as to giving an account of the design and results of the manifestation of God in the flesh, the book is an utter failure, – little better, indeed, than a miserable travesty."[20] What was particularly at issue for the *Patriot* was not necessarily that Christ's moral life was narrated at the exclusion of the facts proving his divinity, but more that the author of *Ecce Homo* had failed to convince that he could in fact be trusted to engage in such a theologically delicate performance. Even though the author's theological views are carefully hidden, argued the *Patriot*, the book shows enough signs that they are in fact "utterly irreconcilable with those that are ordinarily held in the Christian Church, or with acknowledgement of any special deference due to the prophets as the messengers of God."[21] When Macmillan passed the review along to Seeley, he simply referred to it as "stupid" but was surely glad that Seeley's newly published letter to the *Spectator* would implicitly be a reply to the *Patriot* as well.[22]

Unfortunately, as if to contradict entirely Seeley's letter in the *Spectator*, the *British Quarterly Review*'s January article in seeming praise of *Ecce Homo* argued that "The whole book reads like meditations of many years, written and re-written, we should think, to ensure the utmost precision and conciseness."[23] Of course, it was Seeley's claim that he had only just begun the investigation and that the book was not the result of many years of reflection, but the investigation itself. What is more, while the *British Quarterly* was less willing than the *Patriot* to make assumptions about the author's theological views, the reviewer argued that any reader of the book would wonder at the "deductions which the author means himself to draw from his own premises." The reviewer argued that those deductions were largely unpredictable. "It appears to us that they *may* prove to be not incompatible with Anglican orthodoxy, nor is there anything positively to show that they would advance beyond a mere Humanitarianism."[24] While acknowledging that *Ecce Homo* "has rendered noble service to the cause of practical Christian Ethics," the *British Quarterly* was confused by "the views of the writer as to the historic realization of this ideal society," views that are at once "exacting, and yet so vague, as to leave one in deep perplexity."[25] This was not exactly the kind of endorsement of Seeley's possible theology that Macmillan and Seeley were hoping for. The fact that his theology was viewed as indeterminate and possibly unorthodox meant that the *British Quarterly Review*'s opinion was not far off that of the *Patriot*. Macmillan wrote to Seeley to that effect, arguing that "the British Quarterly has fallen into the same line as the Patriot," though he felt that it was not "worth bothering yourself about."[26] At this stage, at least, it appeared that Macmillan's strategy for countering the early criticism was decidedly ineffective.

Moreover, a series of reviews appeared throughout January and early February 1866 that were almost entirely dismissive of *Ecce Homo*. The *Reader*, for instance, wrote rather sarcastically that "Throughout the book the writer labours under the impression that he has something new to tell the world about Our Saviour," the evidence being, of course, that there is nothing new. What is worse is that there is, in fact, "no evidence to show that the author is at all acquainted with the details of the foundation of the Christian commonwealth." The reviewer then lamented the two pages that he had devoted to the review of a book that includes "no trace of the scholar or the theologian in its pages."[27] The *Church and State Review* argued "that its [*Ecce Homo*'s] main spirit is in unison with that of Renan." More problematically, the reviewer found Seeley's statement that no theological questions would be discussed in the book entirely false, that the "whole drift and argument involve theological questions of the very highest possible import."[28]

Macmillan and Seeley do not appear to have been too concerned with the tone of these reviews, no doubt in part because at the same time they were also receiving promising sales notifications. Just three months after *Ecce Homo* had originally hit the bookshops, sales were beginning to take off. On 13 February, a second run of fifteen hundred copies was ordered to be printed.[29] Just a week later, Macmillan reported to Seeley that the second edition had almost sold out,[30] making a much larger print run necessary for the third edition (three thousand copies), to be subscribed in March.[31] And despite some of the expected criticism, Macmillan began to receive praise for the book from some unexpected quarters. Moreover, his various attempts to utilize that support were beginning to pay dividends; he was quietly approaching admirers of *Ecce Homo* to make their praise of the book public, even suggesting possible venues for doing so. We have seen that this strategy backfired with Hutton's review in the *Spectator*, but Macmillan found himself in a position to counter this problem thanks to a rather surprising letter in praise of *Ecce Homo* that he received from one W.E. Gladstone.

The future Prime Minister, who was at the time Chancellor of the Exchequer and a clear rising star in the Liberal Party, wrote to Macmillan that "from the moment when I opened the volume [of *Ecce Homo*], I felt the touch of a powerful hand drawing me on." Gladstone found *Ecce Homo* to be a truly "noble work": "I will venture to say I know of, or recollect, no production of equal force that recent years can boast of, and that it is with infinite relief as well as pleasure that, in the present day, I hail the entrance into the world of a strong constructive book on the Christian system."[32] This was surprising praise because Gladstone was a leading Tractarian, and it was a safe assumption that Catholic-oriented Anglicans would dismiss the book precisely for the methodological division that was proving so unpopular throughout the periodical press.[33] Gladstone did, however, acknowledge that he understood the concern that was being voiced in the periodical press about the seemingly ambiguous theology of the author. It did not help matters that the book was published anonymously, thereby making it difficult for readers to place their trust in an unknown author, who was claiming to engage in an "audacious" journey of historical inquiry. An easy solution would be to give the public the name of the author. "And I venture to add," argued Gladstone, "the opinion that the author of such a work, on such a subject, ought to give the world the benefit also of his name." Such would also no doubt "help to draw to it the attention it deserves, and the more present responsibility of open authorship would be useful, as I believe, even to this remarkable writer, and would make worthier still what I cannot but call this noble book."[34]

Macmillan could hardly contain his excitement upon receiving Gladstone's letter, sending it immediately along to Seeley. "The enclosed from Gladstone is the most important utterance you have called forth yet." (See figure 5.2.) It was certainly necessary to respond to Gladstone's letter, and to do so carefully. Macmillan wanted particularly to underscore his position respecting the book's anonymous status, stressing the need "to rigidly maintain it in the mean time," as he wrote to Seeley. "My formula is that the author has very sufficient grounds for wanting to maintain anonymous at present." Equally important, however, was finding a way to publicize Gladstone's support, not least because it would help *Ecce Homo* find readers who might otherwise simply dismiss it on the basis of previous public criticisms. "I will ask him to let me use his opinion – privately, and if I can see my way to it he might say his say about [it] in some public way. In any case his opinion is sure to be pretty freely given."[35]

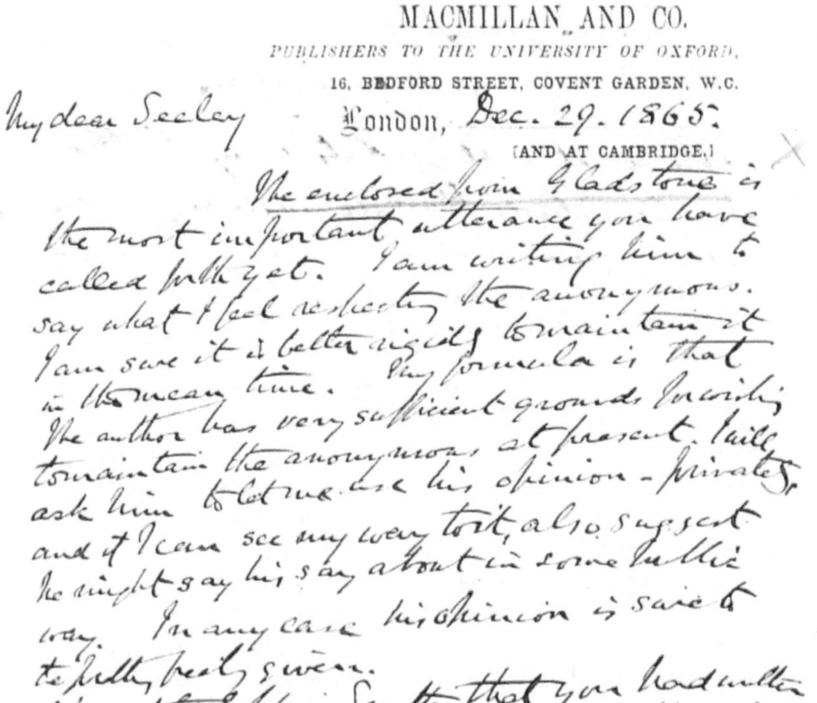

MACMILLAN AND CO.
PUBLISHERS TO THE UNIVERSITY OF OXFORD,
16, BEDFORD STREET, COVENT GARDEN, W.C.

My dear Seeley

London, Dec. 29. 1865.

[AND AT CAMBRIDGE.]

The enclosed from Gladstone is the most important utterance you have called forth yet. I am writing him to say what I feel respecting the anonymous. I am sure it is better rigidly to maintain it in the mean time. My formula is that the author has very sufficient grounds for wishing to maintain the anonymous at present. I will ask him to let me use his opinion – privately and if I can see my way to it, also suggest he might say his say about in some public way. In any case his opinion is sure to be fully freely given.

I have told Edwin Smith that you had written to the Editor of the Spectator yourself. That he would see what you say in the present number probably, and that then if he saw fit he might write. But I again urged even a brief notice in the Daily News.

I have sold two today here – one to Dr Woolley who is on his way to Australia, and one to young Thompson, who contested South Lancashire as Gladstone's colleague. He is on his way to the Southern States.

I am sorry you don't feel well enough to come

Figure 5.2 Macmillan to Seeley, 29 December 1865, Seeley Papers, MS 903/3A/1. Courtesy of the Senate House Library, University of London.

Macmillan was therefore very careful in courting Gladstone's support. In his response, he stressed that anonymity – and remaining anonymous for the time being – was actually an important part of the author's methodology. "Whatever benefit, if any, might come to the book from his name, I think he was moved, beyond the personal considerations which were weighty, by a desire to see the effect on men's minds of the thought that had seemed important to himself unbiased by his reputation, or lack of reputation." In the same way that the author avoided discussing theological matters, he kept back his name to avoid needless debates based on party affiliations. Macmillan implied that this important methodological rationale was missing from the previous reviews of the book, particularly the reviews that appeared in the *Spectator* and the *Patriot*. "Both while speaking highly of the power and originality of the work, blame the writer for doing what he specifically says he did purposefully and which indeed is the very essence and worth of the book … namely, to see what the result of an investigation in a strictly scientific way of this great human … character would be." Macmillan claimed that he was less worried about the *Patriot*, for it "is a dissenting paper of no very great power but on the whole candid and fair." But "The Spectator on the other hand" was another problem entirely, given that it is "perhaps the ablest and most influential of the weeklies." "It would be of great value to the affection and true appreciation of the work," Macmillan continued, "if we could in any way make known the estimate formed of it by distinguished and earnest men." And aside from his original letter of support for *Ecce Homo*, Macmillan assured Gladstone that he "had letters from two eminent Oxford men in quite the same tone." Unfortunately, the *Spectator* review threatened to undermine both the groundswell of support for the book and the book's wonderful message, which "surely Christians should recognise … as most excellent work done for their cause." Macmillan did not explicitly ask Gladstone to write a review, nor did he request that Gladstone find an able mind capable of writing one, though this was certainly the gist of the letter. "I wish much a really good article could be got into the Guardian."[36]

Macmillan and Seeley no doubt would have been keen to have a positive review appear in the *Guardian* because it was an influential weekly with a large circulation, much like the *Spectator*, but also because it was a leading Anglican paper, one that was highly influential among Anglo-Catholic readers.[37] It was edited by the foremost Tractarian, R.W. Church, who was, at the height of the Oxford Movement in the 1840s, one of John Henry Newman's closest confidants.[38] After Newman converted to Catholicism in 1846, Church remained an Anglican, and along with James Mozley and Frederick Rogers founded the *Guardian* to give voice to a broader Tractarianism that was explicitly loyal to the established Church. A carefully crafted review in the *Guardian*, therefore, could work to counter the misleading *Spectator* review while helping to expand *Ecce Homo*'s appeal with an unlikely readership. It was also well known that Gladstone had some influence at the *Guardian*, as he was a good friend of Church and could in all likelihood orchestrate a positive review.

In his response, however, Gladstone did not make any mention of writing a possible review. He once again acknowledged his appreciation of the book, writing that "I am glad to find that it is forcing its way through the clouds of the Anonymous, and will and must be visible to the world." But he admitted that he could understand why the theological ambiguity of the book was so troubling to some readers. "The theology of the work, or of the writer, may, when it appears, for all I know, be very different from what ... God has given the world," Gladstone argued, emphasizing a point made by both the *Patriot* and the *British Quarterly Review*. However, unlike the *British Quarterly*, Gladstone was willing to give the author the benefit of the doubt because "[h]e has built up a broad and solid fabric from the earth towards heaven" and has made "more and more comprehensible the human work of the Redeemer." Somewhat surprisingly, Gladstone was much less supportive in commenting on Seeley's "Letter to the Editor of the 'Spectator,'" which he found needless. "[I]t jarred more than any portion of the book."[39] This was clearly not the effect Macmillan and Seeley sought when responding to the *Spectator* review.

Sensing that he might lose a possible ally, Seeley decided to write to Gladstone himself. Unfortunately, only a draft of this letter remains extant, which was enclosed along with a note from Macmillan which states "*Your letter to Gladstone* is very good on the whole, but I think there are one or two points in which I think it might be improved. See whether my pencil suggestions are worth notice."[40] The fact that the letter was sent to Macmillan for further commentary gives some indication as to how important Macmillan and Seeley believed it to be in maintaining Gladstone's support. And given that Macmillan's suggested revisions are themselves quite telling, the draft of this letter is worth quoting at length. (Macmillan's commentary is indicated in square brackets.)

After graciously thanking Gladstone for his "timely praise," which Macmillan suggested revising to read "sympathy or approval," Seeley wrote that he was grateful that Gladstone was less concerned about the future volume than many other readers. He explained that "the present volume is in no way dependent upon the future one" and that "[i]t depends upon no assumptions which still remain to be proved." This was, Seeley maintained, very much an "inductive" investigation, yet he understood "how difficult it is for men whom I respect, as it were, to give into my hands temporarily all their most cherished beliefs even though I give them back some as I hope, strengthened and enlivened, to value properly what they gain, so long as they are in suspense about what they may yet lose." He claimed that he was particularly "obliged" to Gladstone "for the unreserved cordiality with which you accept" the findings of *Ecce Homo* without knowing the results of the future study. Seeley assured Gladstone that his "results so far are such as no Christian need be alarmed at" but that his method required that he let the evidence guide him regardless of the established religious "opinions of wise men." "But I can promise nothing for I know nothing."

Clearly the main reason for writing to Gladstone, however, had to do with explaining the rationale for the letter in the *Spectator*. "I am sincerely obliged to you for your candid opinion about my letter in the Spectator," Seeley wrote. "I can easily understand that it 'jarred.'" Here Seeley would employ an oft relied upon strategy that was used to explain why the tone of *Ecce Homo* might have come across as objectionable for convinced Christians. As Seeley explained, "I hope you will consider for whom I principally write. It is for those upon whom such words do not jar." Seeley was referring to "the able young men of the day almost without exception [Macmillan struck out "almost without exception," arguing that it was "too sweeping," to be replaced with "the great majority"], with whom the only question in which of the three epithets used in that letter – I will not pain you by repeating them – describe Christ best." The three epithets used in the letter referred to Christ as either a "myth," "deceiver," or "bewildered enthusiast." "I do not wish to shock Christians unnecessarily [struck out with the words "you don't at all"]," Seeley wrote, "but I wrote for those who are not Christians and I should at once lose their attention if I did not show that I am quite prepared to contemplate calmly and consider dispassionately the most extreme unchristian hypotheses." Seeley was adamant that the letter was not "needless," as Gladstone claimed. "Needless for yourself no doubt it probably ["probably" struck out] was, but so intelligent a writer as Mr Hutton had quite misunderstood me and a reviewer in the Patriot still more completely." Seeley wanted to show that his "method is purely inductive ... and that the reader throughout is presumed to be without prepossessions." Seeley went on to say that "Those who refuse to divest themselves of prepossessions are not to be blamed; often no doubt they are quite right; but they have no right to complain of my tone, for I did not write for them, and of this my preface gives them fair warning." Macmillan, perhaps quite rightly, suggested that Seeley delete this last sentence, which was "needless and defiant."[41] The letter was signed from "The Author of Ecce Homo."

It did not take long for Macmillan and Seeley to realize that their courting of Gladstone's favour actually worked. On 7 February a review appeared in the *Guardian*'s pages, and Macmillan had a "strong suspicion that we owe to Gladstone at least the inspiration" for the review, if not authorship of "the article itself."[42] Indeed, the review seemed to follow up on the conversation between Gladstone, Macmillan, and Seeley, even appropriating some of the strategies employed by Macmillan and Seeley in defending the work against critics of the author's methodology. After suggesting that the book was a dangerous one to review, given the fact that there was no name attached to it, the reviewer admitted that "[m]uch of what is on the surface and much of what is inherent in the nature of the work will *jar* painfully on many minds" (emphasis added). But, the reviewer continued, "the subject is put before us in so unusual a way ... [that] a man feels as he does when he is in the presence of something utterly unfamiliar and unique, when common rules and inferences fail him, and in pronouncing upon which he must make something of a venture."[43] There was, in other words, something unspeakably beautiful about *Ecce Homo* that transcended its problems of tone and authorship.

The reviewer therefore sought to calm the nerves of anxious readers about the author's intentions and identity, which the reviewer argued could be better understood by closely reading *Ecce Homo*'s preface as well as the author's letter to the editor of the *Spectator*. The *Guardian* reviewer helpfully included both of these items in full and stressed in particular the notion that the book was not written after the investigation was completed, but was the investigation itself. The reviewer found it commendable that the author of *Ecce Homo* found it his duty to plunge into such difficult questions about the history of Christianity without knowing the answers in advance, and was willing to accept those answers no matter how they might contradict his previous beliefs. This was, indeed, an "audacious and perilous" investigation. But the really difficult question for the reader of *Ecce Homo,* argued the reviewer, was in trying to find out just what was the author's state of mind before engaging in this investigation. Was the author going through a crisis of faith or a crisis of doubt? Just who was the author? It was best not to dwell on this last question, argued the reviewer, because whoever the author might be, he is clearly able "to sympathise with what he sees in the opposing camp. If he is what is called a Liberal, his whole heart is yet pouring itself forth towards the great truths of Christianity. If he is what is called orthodox, his whole intellect is alive to the right and duty of freedom of thought." Indeed, there is a "deep tone of religious seriousness which pervades the work ... Whatever else the book may be, this much is plain on the face of it – it is the work of a mind of extreme originality, depth, refinement, and power; and it is also the work of a religious man: Thomas à Kempis had not a more solemn sense of things unseen and of what is meant by the Imitation of Christ."[44]

The review continued, arguing that *Ecce Homo* is not, as some have said, a work of biblical criticism. It does not attempt to critique the particular facts of the New Testament but rather present the character of Christ that appears by taking into account the Gospels as a whole. The author therefore "answers Strauss as he answers Renan, by producing the interpretation of a character, so living, so in accordance with all before and after, that it overpowers and sweeps away objections; a picture, an analysis or outline if he pleases, which justifies itself and is its own evidence, by its originality and internal consistency."[45] But the work also does not attempt to "turn Christianity into a sentiment or a philosophy." *Ecce Homo* does indeed assert Christianity as "a historical religion and a historical Church; but it also seeks in a manner equally remarkable, to raise and elevate the thoughts of all, on all sides, about Christ, as He showed Himself in the world, and about what Christianity was meant to be." This was, in other words, a book that no Christian should rightly fear, whether liberal or conservative, high or low, because ultimately it asserts "Faith in a Divine Person who is worthy of it, allegiance to a Divine Society which He founded, and union of the hearts in the object for which He created it."[46]

This was the first genuinely positive review to appear in the press. And it also touched on all the right talking points, thereby acting as a response to some of the more prevalent criticisms, and a proxy for the still hidden author. Macmillan was clearly thrilled. "It seems to me admirably done," he wrote to Seeley, "perhaps a little too long, but it puts us on the right rails after the miss of the Spectator … Nothing could be more opportune. It carries you just into the region where your chief danger lay [but you pass by it] with flying colours."[47] Both Seeley and Macmillan would remember the great favour done for the book by their new orthodox friends at the *Guardian*. And just a few days after the review appeared, Macmillan's suspicions about Gladstone's involvement would be proven correct. "I must tell you," Macmillan wrote to Seeley, "the editor of the Guardian told me in confidence that another and very different article was written and in type when some[one] … wrote offering another and calling special and favourable attention to the book." This must have been "our high friend," argued Macmillan. "You remember in my letter [to Gladstone] I named the Guardian."[48] Gladstone must have convinced Church to write the review. And given some of the language that appeared throughout, Gladstone had some input, either directly or indirectly, in what was said. Clearly Church did not need much convincing, given the review's wholehearted endorsement of *Ecce Homo*. (As we will see in chapter 8, however, Church was fairly certain that he knew who the author was, though he was quite wrong.)

The *Guardian* review was certainly cause for celebration. Not only did it mean that a positive review was going to find its way into the hands of thousands of orthodox Anglican readers, it also put forward a message about the book, specifically concerning the author's methodology and intentions, that both Seeley and Macmillan were promoting behind the scenes as it were. "I imagine it [*Ecce Homo*] will be infinitely more palatable to the English religious world than Renan's book was," Matthew Arnold wrote to his mother, "indeed the review in the Guardian may be taken, I suppose, as proof of this."[49] Indeed, the *Guardian* review rather suddenly widened *Ecce Homo*'s possible readership by suggesting that it need not offend the orthodox. Before the review appeared, Macmillan and Seeley had thought about inserting a preface into a new edition of the book, in order to face some of the criticisms directly. This was now deemed unnecessary. The *Guardian* review further justified that decision. "I am glad now," Macmillan wrote to Seeley in thinking about the wider implications of the review, that "you have no preface in your new edition."[50] This issue would be revisited, however, as there would be several new editions on the horizon and many more misinterpretations to correct.

Moreover, for the time being, the close relationship between Seeley and Macmillan also seemed to benefit the reception of the book; they worked well together to coordinate responses to Gladstone in order to counteract the negative press that was surrounding the book. But even at this early stage in *Ecce Homo*'s publishing history, there were signs that publisher and author had different understandings about what was at stake in publishing it anonymously. Just a few days before the *Guardian* review appeared, Macmillan jokingly responded to a query from Seeley about what his father might know about his authorship: "I called on your father yesterday. He [has] no more idea of you being the author than King Kamrani has, who studies Peurch on the margin of the Albert Lake, in a robe of brown bark." Then he suggested sending R.B. Seeley a copy of the book: "I want to send him a copy. May I?"[51] Seeley's response has not survived. But we will see that he was deeply troubled by what his evangelical family might think about him authoring such a book. And as the reception of the book became much nastier in the coming months, this was a growing concern for the author but not so much for the publisher.

Chapter Six

Vomited from the Jaws of Hell

After this crushing exposure, the author will, no doubt, preserve his *incognito*, and we trust the critics of our religious contemporaries will be a little more careful about praising books which they rightly say are "dangerous to review."

– "The Quarterly Review," *Reader*, 14 April 1866

Macmillan was certainly right that the *Guardian* review would have an effect on the future reception of *Ecce Homo*, but it is unlikely that he would have anticipated just how extreme the response would be. Indeed, while the *Guardian* review indicated that *Ecce Homo* was finding some rather unexpected admirers, namely leading Tractarians such as Gladstone and Church, the fact of its very existence meant that conservative churchmen could no longer ignore or dismiss it out of hand. Macmillan helped ensure that this would be the case, as he immediately published an excerpt from the *Guardian* review in new advertisements for the book that made central the *Guardian*'s claim that it was a "dangerous book."[1] The advertisements themselves became the subject of much commentary as well.

This was most apparent in a review that was published in *John Bull*, a strong High Church voice, which found it irksome that *Ecce Homo* "should flaunt its title in advertising columns with the appended *imprimatur* of a respectable Church journal." The "respectable Church journal" referenced here was of course the *Guardian*. *John Bull*, however, was not as easily fooled by the "clever, but extremely objectionable treatise." In attempting to "put himself outside the supernatural atmosphere of … [the] facts and to view them *ab extra* in their purely historical meaning," explained the *John Bull* reviewer, the author of *Ecce Homo* was beginning from a "false position" that could only lead to heterodoxy. The *John Bull* reviewer could not comprehend how the *Guardian* missed this fundamental point. The reviewer wondered what could account for the *Guardian*'s misjudgment, suggesting that perhaps it "has become or is becoming as 'Broad' in its theology as it is 'Liberal' in its politics." This was the only likely explanation for the *Guardian*'s praise for a book that was clearly "written by any one of the more eminent representatives of the Arnold School in our Church, with the ostensible purpose of dragging down or raising up those who soar above or sink below its own nugatory level of *neither-belief-nor-unbelief*, – the *juste-milieu* of anti-dogma."[2]

The thrice-weekly *Record* was equally "astonished at the favourable acceptance the book has received in some quarters where a clearer insight into its doctrinal bearing might have been expected." This was, of course, in reference to the *Guardian*, whose reviewer, much like many other readers, has been taken in by the fact that the book has been "[w]ritten with great subtlety and no little style of beauty." When the book is considered as a whole, however, it is apparent that the author is "no inquirer after truth, whoever he may be," but rather most interested in giving "formal shape to those loose elements of sceptical rationalism." Moreover, as the book is prefaced as a disinterested "investigation," many unfortunate readers fail to grasp this central "insidious and dangerous character" and are thereby led to accept a series of deductions that are supposedly based on this investigation. As the author has "clearly disencumbered himself of all the ancient faith of the Church, and having created for himself a naturalistic scheme of belief in its stead, [he has determined to] lead others down the same dangerous precipice into the same dreary rationalism in which he himself seems to be engulfed." After destroying the doctrinal basis for "the divinity of Christ, the atonement, the person and work of the Spirit, the depravity of human nature, and, underlying them all, the inspiration and authority of the Scriptures," there is nothing left of Christianity beyond a vague notion of "humanitarian perfectibility … The sole hope for suffering human nature is in itself. No grand words can conceal the dreary misery to which mankind is condemned."[3]

While these responses were helping to create a groundswell of negativity from at least one faction of the High Church, who were now claiming that *Ecce Homo* could have been by one of the "eminent representative of the Arnold School" of the English Church, it was just those representatives that were beginning to voice their opinion concerning the book – and they were much less divided than the High Church crowd. The Regius Professor of History at Oxford, Goldwin Smith, for instance, wrote to Macmillan that "I am reading the book over and over with increasing – I will not say admiration – but a much deeper feeling." He also said that he could write a letter to the *Spectator* to help counter Hutton's review, thereby "pointing out the great value and importance of the book in my own way."[4] This was deemed unnecessary when it was decided that Seeley should write to the *Spectator* himself. Macmillan instead "urged even a half notice in the Daily News,"[5] a suggestion that does not seem to have been pursued.

Seeley also received positive commentary from some of the leading figures of liberal Anglicanism, figures Seeley had admired since his time at Christ's College. "Here is Maurice at last," Macmillan wrote to Seeley, enclosing a letter from one of the two most famous "Broad Church divines," F.D. Maurice. "I certainly did not expect the appearance of so striking and original a book in our own day," Maurice wrote, "even less the appearance of one combining so much reverence with such entire freedom of thought." While Maurice admitted that the book begins from the "opposite point to that from which my own thoughts commence, and that our phraseology and modes of contemplating every part of Christ's life are markedly dissimilar," he received much welcome instruction. The author should consider it a "triumph" for "vanquish[ing] the obstinacy and prejudice of an old theologian."[6] This must have been welcome praise for Seeley to receive from a man he so clearly admired, whose theological writings he defended in debates with his father when he was still a young man. "What can be more satisfying?"[7] Macmillan wondered. Just a month later, Macmillan received a letter from the other Broad Church divine, A.P. Stanley, which included some "acute and valuable" criticism that Macmillan assumed would "please and interest" Seeley.[8] Stanley would later publish his critique in *Macmillan's Magazine* (discussed in chapter 8).

Aside from the rather surprising letter from Gladstone and the two from Maurice and Stanley, Macmillan received a host of letters from a variety of individuals connected with British intellectual life. The Dean of St Paul's, Henry Hart Milman, who was a pioneer in introducing German historicism into England, wrote that he read the "remarkable work" with "great satisfaction and pleasure."[9] A man admitting to being a complete stranger, Spencer R. Drummond, wrote to explain that "I cannot find language, sufficiently expressive, to convey to you the thoughts to which the perusal of the 'Ecce Homo' have [*sic*] given rise," though he would try. Even though he had been for fifty-three years engaged in "ministerial thought," he was struck by the "intensely soul-absorbing way" that Seeley permitted "a deeper insight" into "the many subjects upon which you treat."[10] Charles Kingsley's wife, Frances, and eldest daughter, Rose, each wrote to Macmillan expressing their profound appreciation for the book and its veiled author. *Ecce Homo* was "the most important book I ever read except the bible," Frances Kingsley proclaimed.[11] Henry Sidgwick wrote to Macmillan to confess that he had not yet read *Ecce Homo* but that he was going to in a few days in order to see for himself just what was causing such great interest among his peers. "There is a mere chorus of praise about it here [in Cambridge]."[12] Indeed, by February 1866, just three months after the book was first published, Macmillan informed Seeley of "a great excitement and curiosity about the book."[13] And he humorously mused about the state of Seeley's mental health, given the book's sudden and surprising success. "How are you? Does the incense round your invisible throne disturb your brain. I will be the slave to remind you daily that you are mortal."[14] As the second edition quickly sold out in favour of a third, Macmillan was also astounded by the book's success. "The sales go on still," he wrote to Seeley on 15 March, a phrase that would be repeated in the weeks and months to come.[15]

The continuing success of the book was itself becoming a topic of conversation and became a central feature of Macmillan's marketing strategy. Macmillan took out advertising space in all the major weeklies such as the *Athenaeum*, the *Reader*, and the *Spectator* in order to announce new editions of *Ecce Homo*, which were appearing almost every month, along with the sales figures, perhaps as a way to indicate just what a literary phenomena *Ecce Homo* had become. By mid-April, Macmillan was advertising that six thousand copies had been sold.[16] "On the whole it is not very bad to have made £1000 in this quiet way," Macmillan wrote to Seeley about his growing royalties. "I suspect you will make a good deal more yet."[17] And Seeley certainly would, though his time for earning royalties quietly would indeed be over.

As the pace of the book sales picked up and more and more interested parties took notice, conservative opposition to the book began to increase, and in voicing strong denunciations of the book, they of course helped to increase its sales. Indeed, *Ecce Homo*'s growing popularity was engendering a loud and large opposition that was prone to hyperbole, making the book a must-read precisely because of the warnings against it. The first such warning came in early April. Macmillan wrote to Seeley on 3 April 1866, with a "bundle more documents" enclosed with his letter, that "[t]here will be literature gathered around the book much bigger than itself." He flagged for Seeley a letter to the editor of the Anglican *Churchman* as "the most important" of the enclosures.

This was a letter written by George Denison, a Tractarian who first gained notoriety in 1853 when he preached a series of sermons that seemed to defend the Catholic belief of transubstantiation. Though his position was supported by other Tractarians such as R.I. Wilberforce and John Keble, he was sentenced in 1856 to deprivation by the ecclesiastical courts, given that the Thirty-Nine Articles of faith officially declared that a belief in transubstantiation was "repugnant to the plain words of Scripture." In 1857, on appeal at the Court of Arches, the decision was reversed on a technicality. Denison was also one of the central figures behind the mobilization in 1863 to censure Colenso's *Pentateuch and the Book of Joshua Critically Examined* and in 1864 to condemn *Essays and Reviews*.[18] Macmillan was therefore expecting some sort of response from the archdeacon, given his penchant for public controversies. And the archdeacon did not disappoint.

Denison began his letter by explaining that he avoided reading any of the many notices of *Ecce Homo* so that he could make up his own mind about the book. That opinion, rather bluntly stated, was that *Ecce Homo* "appears to me incomparably the most dangerous book of these times." Denison found that the book had "great intellectual pretensions" along with "great attractions of sentiment and style," but these elements were partly what so frightened him. "All this is combined with so many and so mighty errors that I tremble as I read … I shrink from the book with a shrinking I cannot describe, and which I never remember to have felt before; even in respect of its precursors, 'Essays and Reviews,' and Bishop Colenso's books. There seems to me to be a subtle and insidious power in it, very suited to act upon these times; and which, while it claims to contend for Christ, is not for Christ." Denison was claiming not only that *Ecce Homo* was a descendant of *Essays and Reviews* and Colenso's works but also that it was in some ways more dangerous because of its "seduction of sentiment and style." Denison argued that it was therefore the duty of all churchmen to point out the contradictions in the book along with its true anti-Christian meaning in order "to make the reader pause before he allows himself to be led away by the attractiveness of intellectual power, and sentiment, and language."[19] The letter was also reprinted in full in *John Bull*.[20]

Denison's letter indicated a radical turn in the opposition to *Ecce Homo*. It was no longer a book that should simply be ignored or even criticized but, in fact, one that should be feared. Denison trembled for the effect *Ecce Homo* might have on more easily persuaded readers. He believed it to be nothing less than an attempt to undermine the Anglican faith, an attack on the Church itself, and that the Church therefore had to be defended appropriately. This was certainly how Macmillan read "the important move [by] Archdeacon Denison," writing to Seeley that "the war note is evidently sounded. The big dog has bayed his warning," Macmillan continued, meaning that his pack "will yelp in tune."[21]

As if on cue, just a few days after Denison's letter appeared, the Tory *Quarterly Review* published its review of *Ecce Homo*. The *Quarterly* reviewer was at pains to find anything of value in the book. "Apart from the affection of originality, the only novelties we have been able to detect are rash assertions, mistaken principles, and bad taste." The reviewer went on to say that, "judged by its intrinsic merits," the book did not warrant a review in the journal's pages, given that it "appeared to us unworthy to be distinguished from the common run of erroneous books." However, it clearly had a following, one that could count more than a few conservative churchmen among its growing numbers. It was for this reason, in particular, that a review from the voice of conservative orthodoxy was necessary: "the thoughtless approbation which has been bestowed upon [*Ecce Homo*] by orthodox persons is our sole inducement to examine briefly its claim to be accepted by members of the Church of England for a guide to the character and precepts of our Lord."

As this makes clear, the *Quarterly* review was particularly anxious that several orthodox churchmen had at this stage already come out in support of *Ecce Homo*. This could only be explained by the fact that the author had carefully concealed his true beliefs throughout a book that was at best one of shoddy scholarship. "Whatever may be his creed – which he has carefully concealed – his want of candour in dealing with his authorities, his presumption, and his rashness, deserve the severest censure."[22] This review was clearly written in much the same trembling hand as was Denison's letter in the *Churchman*. So concerned was the *Quarterly* that the editor even took out advertising space in some of the weeklies alerting readers to the *Quarterly*'s review of *Ecce Homo* as a "warning article."[23]

Unsurprisingly, given the general tone of the review, it did not take long to get noticed. Just a few days after the April issue appeared, the *Reader* published a brief summary, and was unable to contain its pleasure in the fact that its three-month-old review was vindicated by the *Quarterly*'s takedown. "After this crushing exposure, the author will, no doubt, preserve his *incognito*, and we trust the critics of our religious contemporaries will be a little more careful about praising books which they rightly say are 'dangerous to review.'"[24] This last comment, of course, was in reference to the *Guardian*'s review and the line Macmillan quoted in advertisements for new editions of the book. The message was clear: not only was *Ecce Homo* a dangerous book, it was also one that conservative churchman should be united in condemning. *John Bull* was also delighted by the *Quarterly*'s review: "we are glad to see that our opinion of [*Ecce Homo*'s] dangerous tendency is endorsed by the reviewer."[25]

As far as Macmillan was concerned, the growing High Church opposition could work in the book's favour. He wrote to Seeley to that effect, suggesting that the *Quarterly Review* article would certainly not hurt the book. This was in part because Macmillan, like many others, assumed that the *Quarterly* reviewer was one "Soapy Sam" Wilberforce, the same Bishop of Oxford who had led the charge in the *Quarterly*'s pages in condemning *Essays and Reviews* and *On the Origin of Species* (see chapter 1). If the assumption was that Wilberforce was now leading the opposition against *Ecce Homo* in much the same spirit, it could only mean more publicity for the book itself. "The Q.R. which I read carefully last night is stronger in spite than in anything else," Macmillan wrote to Seeley. "I hear it is the Bp of Oxford ... [I]t has the nasal twang which characterises the true bishop in common with the ranter. I don't think it will hurt."[26] Quite the contrary, Macmillan began to realize that the review might even help.

No doubt this was because Macmillan was receiving letters in support of *Ecce Homo* that were written in direct response to the tone of the *Quarterly* review. W.F. Pollock wrote to Macmillan that the "The article in Q.R. is simply *'brutal'* – it is very unworthy of its reputed authorship ... It is no review, but an attempt to put down the book by force of authority – such cannot succeed in these times and in this country. All thinking people are shocked by it."[27] This last letter was particularly encouraging, since Pollock enclosed it with another letter from his father, Jonathan Frederick Pollock, who was at the time the Chief Baron of the Exchequer, a post he held until 1868. The Chief Baron had apparently read *Ecce Homo* with great interest and was also angered by the *Quarterly*'s response. While the letter from the Chief Baron was returned to Pollock, Macmillan argued that it was significant in showing "that our footing among the orthodox was so far established, before their meaner hounds began to yelp."[28] And in contrast to the barely contained joy in promoting the review that was to be found in *John Bull* and the *Reader*, others pointed out that the *Quarterly*'s review would simply draw more attention to the book, which surely could not have been the reviewer's intention. The *Journal of Sacred Literature and Biblical Record*, for instance, referred to the review as a "violent attack," one that "must do more to increase the popularity of and circulation of the work than any number of favourable notices."[29]

Thus encouraged, Macmillan clearly grasped this point and immediately began running advertisements once the *Quarterly*'s review appeared, announcing the printing of six thousand copies of *Ecce Homo*. As in previous advertisements of this kind, Macmillan included extracts of some of the more positive reviews. This time, however, the first excerpt was from the "*Quarterly Review* on 'Ecce Homo,'" which read: "The shallowest theories and the flimsiest arguments find a ready reception in an empty mind, and their sole strength is in the weakness and credulity of their dupes."[30] This led to a minor controversy between Macmillan and the *Quarterly Review*, as the *Quarterly*'s editor quite rightly believed that the meaning of his review was misrepresented by the carefully selected and misleading quotation.[31] A week later, in an advertisement announcing "seven thousand" copies printed, Macmillan responded by including another sentence in the *Quarterly* excerpt, clarifying the fact that it was a decidedly negative review. The excerpt now began "To refute all the errors which abound in 'Ecce Homo' would be tedious ... The shallowest theories ..."[32] This, I think, goes even beyond the adage that any press is good press. Macmillan was clearly happy to use the fearmongering of the *Quarterly* as a way to help generate interest in the book.

And the *Quarterly Review* seemed only too willing to help out in this regard. In the *Quarterly*'s advertisement for its "warning article" on *Ecce Homo*, an "*ungarbled* extract from the review" was included that essentially filled in the ellipses that separated the sentences quoted in Macmillan's advertisement. In a delicious irony, the *Quarterly*'s advertisement appeared side-by-side with one announcing the printing of "the eighth thousand of ECCE HOMO," thereby providing a wonderful visualization of how the fearmongering of the *Quarterly* could only be helping *Ecce Homo*'s sales.[33] (See figure 6.1.) What is more, given that the *Quarterly* was now regularly advertising a very long excerpt from its review of *Ecce Homo*,[34] Macmillan felt that it was no longer necessary to do so as well, utilizing instead a more favourable excerpt from the *Contemporary Review*.[35]

May 12, 1866.] THE SPECTATOR. 533

ECCE HOMO.

THE

QUARTERLY REVIEW,

No. 238, just published, contains

A WARNING ARTICLE ON THE ABOVE WORK.

The following is an *ungarbled* extract from the Review:—

"To refute all the errors which abound in 'Ecce Homo' would be tedious and useless.

"*Our object is to show the character of the work. The author claims to have studied the subject with especial regard to the facts, and he perverts the commonest particulars which lie on the surface of the Gospels. He writes with an affectation of philosophical depth, and numerous passages in his treatise exhibit either ignorance or defiance of the elementary principles which are familiar to children and peasants. He disguises every-day truths by a pomp of disquisition and a wordiness of style which darken what is simple instead of elucidating what is obscure. His diffuse phraseology is wanting in precision, and his ideas are often in the last degree vague, and sometimes contradictory. His performance is just the reverse of its pretensions, and is inaccurate, superficial, and unsound. Whatever may be his creed—which he has carefully concealed—his want of candour in dealing with his authorities, his presumption, and his rashness, deserve the severest censure. That his book should have obtained the suffrages of any members of the Church of England is melancholy evidence of their slight acquaintance with their faith and their Bibles.*

"The shallowest theories and the flimsiest arguments find a ready reception in an empty mind, and their sole strength is in the weakness and credulity of their dupes. Happily there is a vast body of educated men who are better informed.

"*And while error is perpetually changing its form and is only born to die, the grand truths of Christianity are passed on with accelerated impulse from generation to generation.*"—Quarterly Review, No. 238, pp. 529.

*** The passages in *italics* are omitted by the publishers in the extract from the Review attached to their advertisement of "Ecce Homo."

The same Number of the QUARTERLY REVIEW contains the following articles:—

1—Sir JOSHUA REYNOLDS.
2—CHILDREN'S EMPLOYMENT COMMISSION.
3—FOSS'S JUDGES of ENGLAND.
4—COAL and SMOKE.
5—SCIENCE of LANGUAGE.
6—The IRISH CHURCH.
7—FEMALE EDUCATION.
8—GOVERNMENT REFORM BILL.

JOHN MURRAY, Albemarle street.

THE NEW NOVELS.

Sir OWEN FAIRFAX. By Lady EMILY PONSONBY, Author of "The Discipline of Life," &c.

BOUND to the WHEEL. By John SAUNDERS, Author of "Abel Drake's Wife," &c. 3 vols.

HESTER'S SACRIFICE. By the Author of "St. Olave's," &c. 3 vols.

A NOBLE LIFE. By the Author of "John Halifax," "Christian's Mistake," &c. 2 vols.

MIRK ABBEY. By the Author of "Lost Sir Massingberd," &c. 3 vols. [May 18.

HURST and BLACKETT, 13 Great Marlborough street.

THE POPULAR NOVELS

AT ALL LIBRARIES.

PLAIN JOHN ORPINGTON. By the Author of "Lord Lynn's Wife" and "Lady Flavia." In 3 vols.

"Perhaps the cleverest novel of the class to which it belongs since the publication of 'Lady Audley's Secret.'" —*Globe.*

The ROMANCE of a COURT. 3 vols.

The HIDDEN SIN. 3 vols.

DION and the SIBYLS. By Miles GERALD KEON, Colonial Secretary, Bermuda. 2 vols. post 8vo.

"We do not remember anything of the kind which can stand between it and 'The Last Days of Pompeii.'"—*Reader.*

RICHARD BENTLEY, New Burlington street.

ALLEN'S INDIAN MAIL and OFFICIAL GAZETTE.—Latest News from all parts of India—Latest Government Appointments—Latest Information regarding the Services—Notes on all Indian topics likely to interest those who have resided in India or have friends there. Published four times a month, on the arrival of the Marseilles Mail from India. Subscription £1 4s. per annum, payable in advance; specimen copy, 6d.

London: WM. H. ALLEN & Co., 13 Waterloo place, S.W.

34th Edition, price 6d. and 1s.

NEUROTONICS; or, the Art of Strengthening the Nerves, containing Remarks on the Influence of the Nervous System upon the Human Economy, with Illustrations of a New Mode of Treatment for Chronic Diseases, Nervousness, Debility, Low Spirits, Indigestion, &c. By D. NAPIER, M.D.

Through any Booksellers; or free for 7 or 13 stamps, from the Author, 14 Allen road, Stoke Newington, N.

THE EIGHTH THOUSAND OF

ECCE HOMO:

A SURVEY OF THE

LIFE AND WORK OF JESUS CHRIST.

8vo, cloth, 10s 6d.

"It would indeed be easy to give colour to very opposite estimates of the general cast of thought of an author whose mind moves so little in the general plane either of dogmatic theology or of modern philosophical criticism. Few books are more capable of being unfairly represented by extracts. Few are less capable of being estimated from a summary or an abstract. The author's whole view of his subject is essentially a layman's view,—practical, almost lawyerlike in the directness with which its points are made out. It is, in short, an application of work-day common sense to a subject ordinarily reserved for technical or devotional treatment. In this lies the secret both of the charm of the book and of its incompleteness. The style is remarkable for clearness, vigour, and simplicity. It has a reality and living force which make even old things seem new, and never leave the reader for an instant in doubt of the exact meaning of what is new........If we have spoken freely of what we think errors and defects in the book it is because we feel that it has at present, and deserves to have, a degree of influence upon the minds of educated and thinking men, such as very few books in any generation can exert."—Rev. E. T. Vaughan, in the *Contemporary Review* on "Ecce Homo."

MACMILLAN and Co., London.

MAX MULLER'S HANDBOOKS for the STUDY of SANSKRIT.

On Friday next, the 18th inst., in royal 8vo.

A SANSKRIT GRAMMAR for BEGINNERS, in Devanagari and Roman Letters throughout. By MAX MULLER, M.A., Taylorian Professor at Oxford.

HITOPADESA, Book I., edited by Prof. Max Muller, with Transliteration, Interlinear Translation, and Grammatical Analysis, 7s 6d. Sanskrit Text only, 3s 6d.

HITOPADESA, Books II., III., and IV., by the same Editor, with Transliteration, &c., 7s 6d. Sanskrit Text only, 3s 6d.

BENFEY'S SANSKRIT-ENGLISH DICTIONARY, in Devanagari and Roman Letters throughout, 8vo, 52s 6d.

London: LONGMANS, GREEN, and Co., Paternoster row.

Now ready, price 5s, the Second Edition, revised, of

CHRISTIANITY without JUDAISM. By the Rev. BADEN POWELL, M.A., F.R.S., late Savilian Professor of Geometry in the University of Oxford.

London: LONGMANS, GREEN, and Co., Paternoster row.

This day, 8vo, 3s.

CONTRIBUTIONS to an INQUIRY into the STATE of IRELAND. By the Right Hon. Lord DUFFERIN, K.P.

JOHN MURRAY, Albemarle street.

Just published, crown 8vo, 7s 6d.

A MANUAL of HUMAN CULTURE. By MICHAEL ANGELO GARVEY, LL.B., Barrister-at-Law (being a concise exposition of the Theory of Education and of its practical application).

London: BELL and DALDY, 186 Fleet street.

Now ready, crown 8vo, price 7s.

THE SONGS and BALLADS of CUMBERLAND; with Biographical Sketches, Notes, Glossary, and Portrait of Miss Blamire. Edited by SIDNEY GILPIN.

London: GEORGE ROUTLEDGE and SONS; Edinburgh: JOHN MENZIES; Carlisle: GEORGE COWARD.

Now ready, demy 8vo, 32 pages, price 4d., free by post 5d.

REPRESENTATION of the PEOPLE BILL. Second Reading. SPEECH of the Right Hon. ROBERT LOWE, M.P., House of Commons, Thursday, April 26, 1866. Revised and corrected.

ROBERT JAMES BUSH, 32 Charing cross.

LONDON LIBRARY, 12 St. James's Square, S.W.

The ANNUAL MEETING of MEMBERS will be held in the Reading-Room on SATURDAY, the 26th inst., at three o'clock in the afternoon.

By Order of the Committee,

ROBERT HARRISON, Secretary and Librarian.

9th May, 1866.

PUBLIC and PRIVATE LIBRARIES and the GENERAL BOOKBUYER. The choice of Books is now offered in a Series of Catalogues just issued—No. 1, comprising a Selection of SOLD-OFF BOOKS and REMAINDERS, published from 1s to £13 13s, now reduced in price from 3d. and so on to £3 3s. No. 2, comprising SURPLUS BOOKS of recent date, many published during last Christmas season, offered at 25 and 30 per cent. discount. No. 3, comprising all the RECENT PURCHASES of valuable illustrated, illuminated, and other books, published from 1s up to £201, now reduced in price to 6d. and so on to £38. Applications to insure the above must be made either personally, or by letter, as none will be sent without, to S. and T. GILBERT, Booksellers, 4 Copthall Buildings (back of the Bank of England), London, E.C.—The above Catalogues gratis and postage free.

Fifth Thousand, price 10s, 2 vols, fcap 8vo, cloth.

A MANUAL of BRITISH BUTTERFLIES and MOTHS. By H. T. STAINTON, F.L.S. Containing descriptions of nearly 2,000 Species, interspersed with "readable matter," and above 200 Woodcuts.

JOHN VAN VOORST, 1 Paternoster row.

NEW PUBLICATIONS.

NOW READY AT ALL THE LIBRARIES

PROFESSOR KINGSLEY'S NEW NOVEL.

HEREWARD THE WAKE,

"Last of the English."

By CHARLES KINGSLEY, M.A.

2 vols. crown 8vo., cloth, price 21s.

THE DOVE IN THE EAGLE'S NEST.

By the Author of "The Heir of Redclyffe."

2 vols. crown 8vo., price 12s.

CLEMENCY FRANKLYN.

By the Author of "Janet's Home."

2 vols. crown 8vo., price 21s.

Mr. HENRY KINGSLEY'S NEW NOVEL.

LEIGHTON COURT:

A Country-House Story.

By HENRY KINGSLEY, Author of "Ravenshoe," "The Hillyars and the Burtons," "Austin Elliot," &c.

2 vols., crown 8vo, cloth, price 21s.

A SON OF THE SOIL.

A New Novel.

2 vols., crown 8vo, price 21s.

With a coloured Map and Illustrations.

ACROSS MEXICO IN 1864-5.

By W. H. BULLOCK.

Crown 8vo., cloth, price 10s 6d.

DUKE ERNEST: a TRAGEDY.

AND OTHER POEMS.

By ROSAMOND HERVEY.

Fcap. 8vo., cloth, price 6s.

ESSAYS ON ART.

By FRANCIS TURNER PALGRAVE, M.A., late Fellow of Exeter College, Oxford.

MULREADY—DYCE—HOLMAN HUNT—HERBERT—POETRY, PROSE, and SENSATIONALISM in ART—SCULPTURE in ENGLAND—the ALBERT CROSS, &c.

Extra fcap. 8vo., cloth, price 6s. (Uniform with "Arnold's Essays.")

Vol. VIII., 8vo., cloth, price 10s 6d, of

THE CAMBRIDGE SHAKESPEARE.

THE WORKS OF WILLIAM SHAKESPEARE.

Edited by W. G. CLARK and W. ALDIS WRIGHT.

CONTENTS: Hamlet—A Reprint of the Edition of 1603, "The Tragicall Historie of Hamlet, Prince of Denmark"—King Lear—and Othello.

Volume IX., completing the Work, is in the press.

MACMILLAN and Co., London.

Shortly will be published.

ST. PAUL'S EPISTLES

TO THE

EPHESIANS, COLOSSIANS, AND PHILEMON.

With Introduction and Notes, and an Essay on the Traces of Foreign Elements in the Theology of these Epistles.

By the Rev. J. LLEWELYN DAVIES, M.A.

Rector of Christ Church, St. Marylebone.

MACMILLAN and Co., London.

This day is published, crown 8vo, cloth, price 7s 6d.

DISCOURSES. By Alexander J. Scott, M.A., Professor of Logic in Owen's College, Manchester.

MACMILLAN and Co., London.

This day is published, the Fifth Edition, crown 8vo, cloth, price 10s 6d.

SERMONS. By the Rev. Henry WOODWARD, M.A., formerly of Corpus Christi College, Oxford, Rector of Fethard, in the Diocese of Cashel.

MACMILLAN and Co., London.

This day is published, crown 8vo, cloth, price 5s.

The HULSEIAN LECTURES for 1865.

OUR LORD JESUS CHRIST the SUBJECT of GROWTH in WISDOM. Four Sermons (being Hulseian Lectures for 1865) preached before the University of Cambridge. To which are added Three Sermons preached before the University of Cambridge in February, 1864. By the Rev. J. MOORHOUSE, M.A., St. John's College.

MACMILLAN and Co., London.

OXFORD on MAY-DAY.—PRINCE CONSORT MEMORIAL.—The BUILDER of THIS DAY, price 4d, by post, 5d, contains:—Fine View and Plan of New Residences, Christ Church College, Oxford—The Brighton Railway Collision—The Exhibition of the Royal Academy—Architecture at the Royal Academy—Exhibition of National Portraits—Progress of the National Prince Consort Memorial in Hyde Park—The Institution of Civil Engineers—Sanitary Matters—Breadth of Light and Shadow in Architecture—Professional Etiquette—The Trades Movement—Thrust of Gothic Arches—Provincial News, &c.—Office, 1 York street, Covent Garden, and all Booksellers.

Figure 6.1 Duelling advertisements in the *Spectator*, 12 May 1866. Courtesy of the *Spectator*.

By May, Denison's pack was still yelping, and this metaphor perhaps best captures Lord Shaftesbury's infamous criticism of *Ecce Homo*, which did not come in the form of a review or even a letter to the editor, but was rather shouted during a public meeting. We have come across Lord Shaftesbury before (chapter 2). Seeley's father was one of Shaftesbury's local advisors and the two men worked together on many evangelical initiatives, most relevantly establishing the Church Pastoral Aid Society. At an early May meeting of that society at Exeter College, Oxford, Shaftesbury explained that he had come across "a great professor of Evangelical religion" who approved of *Ecce Homo* as conferring "great benefit upon his own soul." This should be a cause for deep concern, Shaftesbury argued, because if those apparently well schooled in evangelical thought can be taken in by the literary power of *Ecce Homo*, "how can we expect that the mass of the people, the mass even of the educated middle classes, who are supposed to think for themselves, will not be led to wander out of the right way." It was therefore important to emphasize that "*Ecce Homo* was the most pestilential book ever vomited from the jaws of hell."[36]

Shaftesbury's outburst was widely reported in both the periodical and newspaper presses,[37] and was often ridiculed.[38] The *Saturday Review*, in its typically sarcastic tone, published an article entitled "Lord Shaftesbury's Inferno," in reference to "that subjective hell which exists in Lord Shaftesbury's imagination." The *Saturday* reviewer admitted that Shaftesbury was prone to this kind of hyperbole: "when Lord Shaftesbury informs a gentleman that his pestilential work has been vomited from the jaws of hell, it means that he disagrees with him." Such would be Shaftesbury's complaint about any person who did not conform to his own orthodox opinions. "Considered in this way, no one can hold himself to be seriously aggrieved by having his works traced to 'the jaws of hell.'"

The *Saturday* reviewer continued that Shaftesbury's outburst should best be understood as a "rather feminine exaggeration of speech" that is "characteristic of an antiquated form of controversy," one that was more suited to a bygone age. Indeed, "so long as we were in the habit of burning people by way of hinting our disapprobation of their doctrines, there was nothing shocking about expressing a belief in the most frightful consequences of erroneous opinions." It no longer makes any sense to condemn one's adversaries in such a way. And, what is more, doing so can lead to "unforeseen consequences" that actually favour those being attacked. The reviewer even mentioned the way in which passages from the scathing *Quarterly Review* article were quoted in advertisements for *Ecce Homo* by the book's publisher.

> We need not ask whether they have done wisely; the condemnation of the *Quarterly Review* is certainly of far more weight than this rant about the "jaws of hell"; but the device suggests that some people, at any rate, accept a hostile criticism as at least an incentive to read the book condemned. It would not be the first time that an *Index Expurgatorius* would serve as an excellent advertising medium. Lord Shaftesbury's figure of speech may induce some of his own friends to satisfy their curiosity as to matter vomited from such a singular quarter. But, in any case, the condemnation cannot fail to give pleasure to the writer assailed. For it is plain that it can only mean two things, when we come to analyse the assertion. It means, in the first place, that Lord Shaftesbury disagrees with the book – a fact which can be no news to its author; and, secondly, considering that anything vomited by the jaws of hell is likely to be tolerably destructive, it must mean that Lord Shaftesbury is considerably frightened by the effect which the book is producing – a fact which, so far as it may be new to the author, can be nothing but gratifying.[39]

Peter Bayne, writing for the *Fortnightly Review*, found Shaftesbury's comments rather sad, arguing that the right response to them "is pity for his lordship." Shaftesbury was to be pitied, Bayne argued, because he was so clearly under the influence of such an extreme fear that caused him to lash out at any intrusion into his narrow view of orthodoxy. Much like Denison's, Shaftesbury's response to *Ecce Homo* followed from his similarly hyperbolic criticisms of Colenso and *Essays and Reviews*. His failure to engage critically with these works was evidence that he really did not know what particularly was wrong with them, just that they played a kind of music that "pierces his drowsy ear. What does it mean? He does not in the least know what it means; but, orthodox or heterodox, it is clearly not the old, old song to which he has listened to from his infancy; and therefore it must be infidel, atrocious, cruel."[40]

Shaftesbury himself seemed rather unmoved by the criticism. He noted in his diary that he "denounced 'Ecce Homo' as a 'most pestilential book,'" though he could not quite recall saying "ever vomited from the jaws of hell," as a report had suggested. He did not doubt the accuracy of the report, however. "No doubt, then, I used the words.

They have excited a good deal of wrath. Be it so." He was willing to admit that his precise choice of words was, "perhaps, too strong for the world, but not too strong for the truth. It escaped, in the heat of declamation, justifiable and yet injudicious."[41] Even while speaking at a meeting of the Bible Society at Salisbury in October, Shaftesbury was asked to reconsider his strongly stated opinion about *Ecce Homo*, at which point he claimed that "he could not retract a word of it."[42] He seemed not to realize that the real problem with his chosen words was that they would, as the *Saturday Review* made clear, be used as a free advertisement for the book, much as had the *Quarterly Review*'s denunciation just a few weeks before.

Unsurprisingly, the sales of *Ecce Homo* increased after the intervention of the *Quarterly* and Lord Shaftesbury. But even Macmillan, who sought to utilize their criticisms, was surprised at just how effective they were in helping to sell the book. Just ten days after the *Quarterly*'s review appeared, Macmillan wrote to Seeley that the fourth edition was just printed and that it had "sold over 1200 copies which I confess surprised even me." This brought the overall sales of *Ecce Homo* to seven thousand copies.[43] By the second week of May, Macmillan advertised that eight thousand copies had been sold.[44] And in early June, thanks to "Lord Shaftesbury's inferno," in just ten months *Ecce Homo* was announced as having sold nine thousand copies.[45] Indeed, it seems that many new readers were excited by the prospect of discovering for themselves a book that might have been vomited from the jaws of hell. And this rapid and celebrated interest in the book of course led to an increase in the number of reviews in the weekly, monthly, and periodical presses, which multiplied dramatically throughout the spring and summer of 1866.

In May, June, and July alone, reviews appeared in the *Contemporary Review*, the *London Review*, the *Fortnightly Review*, *Fraser's Magazine*, *Macmillan's Magazine*, the *Month*, the *Quiver*, the *British Foreign and Evangelical Review*, the *Journal of Sacred Literature and Biblical Record*, the *Christian Observer*, the *North American Review*, the *Christian Remembrancer*, the *Dublin Review*, the *Dublin University Magazine*, the *Examiner*, *Sword and Trowel*, and the *Westminster Review*.[46] Immediately following the extreme reactions of the *Quarterly*, Denison, and Shaftesbury, reviewers were anxious to avoid being associated with such ad hominem attacks on *Ecce Homo*, which themselves were becoming the subject of much derision. Even though the reviewer for the *London Review* admitted to "not agreeing in many matters with this unknown writer," the reviewer wanted to be clear that "we are most unwilling to be suspected of joining in an attack upon him for what he has not done." Here the reviewer was explicitly referring to Seeley's method of limiting himself to the Gospels, a method that the reviewer felt was entirely appropriate for a study that was essentially about the relationship between Christ and modern society, that such "human virtues are due in their expanded form to Him." This was certainly a novel study, according to the reviewer, who found the book's critics failing to recognize that novelty itself was an important aspect of Christian renewal. "The author is attacked as an innovator. Those who do so, seem to forget that it is the beauty of the Christian revelation to be constantly new and fresh to each successive student." In this regard, *Ecce Homo* was just the sort of analysis of Jesus Christ that was necessary for the present age: "History, science, all we know, enables us to see more clearly the central source of light, the Sun of Righteousness which alone shone as clearly to our fathers, though they could not see His face through the mists of an ignorant age." The reviewer even believed that it had succeeded in its stated goal by having an "effect ... upon the undecided and the sceptical, who have arisen from it, not with heterodox opinions or diminished faith, but a better appreciation of the great example, and a desire to serve Him of whose divinity 'Ecce Homo' had awakened no doubt."[47]

If *Ecce Homo* was to be criticized, however, it was in the way in which it deifies science, often at the expense of the very religion it claims to be salvaging. "We cannot forbear to add a passing remark on the adhesion here given to the common deification of natural science, to which the author's philosophical bias and admiration of his own age would naturally lead him." In this regard the author "pushes ... the cant of science to a monstrous extreme," particularly when he suggests that contemporary men of science know more about revelation than someone like Moses. "Science may be almost called the lowest kind of revelation," argued the reviewer, "but it is different in kind as well as in degree even from the revelation of moral truth: How far then from the revelation of divine truth!"[48]

Edward Vaughan, writing in the *Contemporary Review*, followed a similar line to that of the *London Review*. He too defended Seeley's method and tone, arguing that the book "lies almost wholly within the province of Christian evidences, not of theology proper," and that the often "jarring" language that is used "is meant for those who are without, not for those who are within the circle of faith."[49] Again, *Ecce*

Homo as a scientific investigation is stressed, as Vaughan points out that "The author places himself beside a supposed inquirer, prepared to work his way (as we believe that he himself has done) toward Christian belief, but determined to take nothing for granted at the outset, but that which cannot be denied without destroying all common ground upon which an investigation can proceed."[50] But even while Vaughan found much to sympathize with in the author's approach, he believed that it led to a single great defect. Because of the scientific and historical approach, "It does not sufficiently recognise the fact of *sin* as one of the universal facts of human nature ... And consequently it does not sufficiently recognise the truth that Jesus Christ, revealed as the Saviour of his people *from their sins*, was the charm which above all drew souls to follow Him."[51]

The *Quiver*, rather surprisingly, followed this general line of critique. It is surprising because it was an illustrated evangelical magazine devoted to the "defence of biblical truth," according to its subtitle. But the reviewer found that *Ecce Homo*'s author was not a radical unbeliever but was, in fact, a true believer "in Christ's Divinity, and the reality of his miracles." The reviewer was also sympathetic to the fragmentary nature of the volume, arguing that the most Christian course of action to pursue would be to "suspend judgment on [the author's] orthodoxy until we have his entire work before us." Moreover, the reviewer claimed to be fairly certain that such a future volume would likely state "in a more categorical manner" the author's belief that Christ's society could only be "satisfactorily ... accounted for by the real essential Divinity of Him who accomplished it." That said, much like Vaughan, the reviewer found one "very serious and dangerous defect." This was the author's rather blinkered "humanistic point of view," which worked to undermine his ability to present Christ as he really appeared to his followers, not just as a moral guide but also as the embodiment of Divine power. Because "the humanistic is the predominating element in his estimate of the character of Jesus," the reviewer believed that "[t]he general tendency of the work is decidedly dangerous."[52]

Even though the *Contemporary Review*, the *London Review*, and the *Quiver* had very different editorial perspectives, their respective reviews of *Ecce Homo* each sought to avoid succumbing to the hyperbole that had dominated the reception of the book up until then. And yet, in their different ways, they also helped reinforce the view that there was something radical about the book, whether it lay in the humanistic perspective of the author or in his adoption of a scientific and historical methodology. Such reviews, along with the competing advertisements referring to *Ecce Homo* as a "dangerous book," engendered a set of expectations for new readers of the book that were difficult for *Ecce Homo* to reach. Indeed, those hoping to read such a novel book that supposedly deified science were often sorely disappointed.

Chapter Seven

A Sheep in Wolf's Clothing

The skin is the skin of the rationalist wolf, but the voice is the voice of the tamer and more orthodox animal.

– "Ecce Homo. First Notice," *Fraser's Magazine* (June 1866)

Ecce Homo was vomited from the jaws of hell! The *Quarterly Review* issued a "warning article" against the book. Denison trembled as he had never trembled before. Even reasonable reviews such as those in the *Quiver* and the *London Review* found that *Ecce Homo* was ultimately dangerous in its general line of thought. And, as we have also seen, Macmillan exploited such views in advertisements for the book under the correct assumption that they would increase its readership. But what Macmillan may not have anticipated was that by advertising *Ecce Homo* as a "dangerous book," he may have led new readers to expect a historical Jesus that would live up to such a description. "Have you seen Ecce Homo?" Major Bell wrote to Mary Elizabeth Roberts on 8 May 1866, both individuals associated with the freethinking *Westminster* crowd. "It is a sort of Life of Christ, which has run thro' eight editions in about three months, and the authorship of which, is still a secret. I don't think it a very remarkable book myself: it seems to me to be timid and inconsistent."[1] Indeed, by the end of May this kind of disappointed response to *Ecce Homo* was forming a distinct trend in its reception.

James Fitzjames Stephen, for instance, writing for *Fraser's Magazine*, clearly had particular expectations when he began to read *Ecce Homo*. He argued that the book was "nearly the first attempt which has been made to write in English a life of Christ like those which have exercised so much influence in France and Germany." The popularity of the book proved to Stephen "what an intense feeling of expectation" the English public has had for just such a book to deal with "those great questions which have hitherto been discussed in this country, as it were, under the breath and by a minority of writers almost infinitesimally small." Despite the fact that England, according to Stephen, is the freest of European countries, intellectually it tends to be much more orthodox than all other countries where free discussion takes place. Stephen claimed that it is the apparent purpose of *Ecce Homo* to embrace the freedom of thought that is seemingly only practised in other countries by inquiring seriously into the facts of the Christian past by relying on "common historical principles."[2]

Stephen argued, however, that while this was precisely how the author of *Ecce Homo* presented himself in the preface, that is as a radical freethinker, he did so in order to conceal his actually quite conservative and orthodox views, which are more thoroughly presented in the main section of the book. "[T]he book is a novel," argued Stephen, "and not a good novel, under a critical disguise. It gives the impression of being written by a sheep in wolf's clothing. The skin is the skin of the rationalist wolf, but the voice is the voice of the tamer and more orthodox animal."[3] The author presents himself as a rationalist who will only follow the logic of the facts, critically weighed, and yet he does something closer to the opposite. In particular, Stephen found the supposedly historical method entirely confused, with no critical analysis of any of the sources upon which the author relies. "Any one who pretends to 'weigh critically' the facts relating to Christ," argued Stephen, "ought to have begun by weighing critically the books from which we may learn something as to the time and people amongst whom he lived and worked." Stephen believed that this central "fault appears to us to destroy altogether the solid value of the book, and to reduce it to the rank of a mere work of imagination."[4] Because of this fundamental failure to appreciate the most basic premise of biblical criticism, "The author of *Ecce Homo* has not proved that he has any right to speak upon such a subject at all."[5] This was damaging criticism indeed.

Even worse, perhaps, were the opinions that were being voiced by Seeley's friend Henry Sidgwick. As discussed in chapter 1, the two men became friends at Cambridge when Seeley was a fellow of Christ's College and Sidgwick was an undergraduate at Trinity College. More specifically, Seeley actually met Sidgwick through the former's brother, Leonard. Along with their mutual friend Frederic Farrar, the four often met and discussed the work of Coleridge, Carlyle, and Maurice.[6]

They were also in agreement about the necessity of establishing the historical evidence for Christianity on sound scientific principles. It was not, therefore, difficult for Sidgwick to come to the conclusion that Seeley was the great unknown author of *Ecce Homo*, as the two of them had often spoken about many of the ideas about Christ's life that appeared in the book. As we will see in chapter 9, Sidgwick's discovery of Seeley's authorship proved to be quite stressful for the still veiled author, but once Sidgwick was let in on the secret he became an important correspondent for Seeley, as the two men were able to engage in a critical but friendly discussion about the merits of *Ecce Homo* that did not have to be mediated by Macmillan.

Sidgwick was initially quite excited about the possibilities offered by *Ecce Homo*. Upon a first read he found that it was a truly "great work" that was "powerful" and "absorbing," though he was ultimately unsatisfied with the author's method "because he passes so lightly over critical questions."[7] After reading and thinking more about the book, he wrote to his friend H.G. Dakyns that he believed "*Ecce Homo* will turn out to be a broken reed, but it is just the kind that *does* not run into a man's hand and pierce it – in fact, I think there will always be a stump left."[8] He was clearly torn by what the volume promised and what it ultimately achieved. He found the analysis of Christ's life truly "sublime"; at the same time he could not help but think that that life was not established by the historical principles promised in the author's preface.[9]

Sidgwick said as much to Seeley in their correspondence about *Ecce Homo*. Much of the problem for Sidgwick hinged on the issue of the historical reality of miracles. Sidgwick found it contradictory that Seeley could exclude the evidentiary basis for some miracles while not also doing so for the non-miraculous facts of Christ's life that were being appropriated from the same biblical sources. "If miracles are not to be believed here (or they are believed no where else) we conclude that Legend has freely handled the historic facts of Jesus' life: then comes the question, what facts narrated are of a kind to have been added or transformed by the same modifying influences that introduced the miraculous element[?]" He continued that "[i]t is unreasonable to assume that all the miracles here have been mistaken and nothing else along with them."[10] In other words, Sidgwick found it difficult to understand how Seeley was choosing between the facts he was relying upon within the same sources. How could he justify relying on a given source for some facts when he had already determined that the source itself was incapable of being relied upon when it came to the miraculous events of

Jesus's life? If the Gospels were not to be trusted on some topics, how could they be trusted at all? Sidgwick, much like Stephen, expected that there would be a discussion explaining how the author was coming to these kinds of evaluative decisions in presenting his evidence, but Sidgwick found that the very short discussion about sources that appears in chapter 5 of *Ecce Homo* had actually worked to further confuse the matter. This became a central point of contention in Sidgwick's review, which appeared anonymously in the freethinking *Westminster Review.*

In the review, Sidgwick was much less confrontational than Stephen, while following a similar line of critique. He presented himself as a friendly reviewer, as someone in general agreement with the main thrust of *Ecce Homo*, namely that "the influence of Jesus on the modern world should increase and not decrease."[11] Moreover, Sidgwick stated at the outset that he had to "give warm and sincere praise to the originality of the conception, the vigour of its execution, the sympathetic intensity with which the writer has grasped the chief points in the character and work of Jesus, the flowing and fervid eloquence with which he has impressed them on his readers." According to Sidgwick, "It requires genius to produce this effect: and genius of a certain kind our author possesses."[12] But this glowing praise quickly shifted to pointed criticisms of Seeley's method, which Sidgwick argued was "radically wrong." He also found *Ecce Homo*'s conclusions "only roughly and partially right."[13]

As was already clear in their correspondence, Sidgwick explained that his main concern with the volume had to do with the evaluation of the evidence that was used as a basis for *Ecce Homo*'s historical claims. Complicating the issue for Sidgwick was the fact that its historical methodology was itself poorly explained. "The first thing that will surprise a student who has taken up the book, is the total absence of any introductory discussion of the evidence on which the historical portion of the book is intended to be based."[14] The only discussion of evidence appears, oddly, in the middle of the book, and what is said there about relying on what is common between the synoptic Gospels is entirely inadequate. This discussion was problematic for Sidgwick for several reasons. First and foremost he argued that "the critical school will hardly admit that all that the synoptic Gospels have in common may be relied upon as certain." Interestingly, whereas many orthodox reviewers argued that *Ecce Homo* was a blind application of the critical school to the history of Christ, Sidgwick found that Seeley actually failed to live up to some of the most basic premises of historical criticism.

This is most apparent in Seeley's brief discussion of Christ's miracles, which in theory Seeley was to avoid, in keeping with his method of divorcing the theological from the historical. But, given that the miracles are recounted in the synoptic Gospels, Seeley's method of accepting that which is common throughout forces him to "provisionally ... speak of them as real." In this instance, as well, according to Sidgwick, we find the author taking an extremely peculiar stance that is seemingly ignorant of recent debates. "To speak of miracles 'provisionally as real' is the one thing that no one will do," Sidgwick argued, whether we are referring to a practitioner of biblical criticism such as David Strauss or a High Church Anglican such as George Denison. "The question of their reality stands at the threshold of the subject, and can by no device be conjured away."[15]

Sidgwick therefore challenged Seeley's argument that he had only accepted the conclusions about the life of Christ based on what "the facts, critically weighted, appear to warrant." Sidgwick believed that Seeley was sincere in attempting to do this, but he argued that "we could not select more appropriate words to describe what ... he has omitted to do."[16] By not adequately discussing his use of the sources and thoroughly explaining why apparent facts are discounted while others are not, Seeley makes some of the most basic threads in his argument, such as Christ's supposed "enthusiasm for humanity," appear suspicious.[17] "The consequence of all this is that the many good things he has to say about Christ's legislation are useless to the accurate reader in their present form, because the framework in which they are placed is so carelessly and clumsily constructed out of unsupported assumptions." And this is really what was problematic for Sidgwick: Seeley was trying to historicize what Victorians all admired in Jesus, but his method made even those facts seem controvertible, thereby engendering an "uneasy feeling that even what we admire in him [Christ] may prove unsound when closely tested."[18]

During their correspondence, it is quite clear that Sidgwick's attempts to come to terms with *Ecce Homo* actually had the opposite effect to what Seeley intended. Sidgwick's faith was not being strengthened but undermined. He even suggested that he might not be able to publish his critique of *Ecce Homo*, as it would likely appear anti-Christian. "I am working myself up gradually to a pitch of irritation with your ideas," Sidgwick wrote to Seeley: "the reason is that I have no outlet, because I do not on the whole feel that I have a right as a fellow of a college to publish anything distinctly antichristian in its tendencies, and any serious criticism of your book that I undertook must come to that."[19] Sidgwick was, in a way, coming to the same conclusions with *Ecce Homo* that Frederic Harrison reached with *Essays and Reviews*: following the logic of the critical method would necessarily lead to an anti-Christian position. And in criticizing *Ecce Homo*, Sidgwick felt that, as a subscriber to the Anglican articles of faith, he could not, legally speaking, make such a claim. He rationalized his own seemingly anti-Christian views by arguing that he could "read my own ideal into the [orthodox] account" of Christ's life.[20] Just a few years later, however, in 1869, Sidgwick determined that he could no longer in good faith hold an appointment that required formal subscription to the articles of faith and resigned his Trinity fellowship.[21]

Whether or not *Ecce Homo* played a direct role in Sidgwick's road to unbelief is unclear. But his response to *Ecce Homo*, which is largely symbolic of the general rationalist reaction to the book, suggested to Seeley and Macmillan that such criticisms would need to be addressed in a new preface. As early as 20 April, Macmillan advised that a new preface should accompany a new edition of the book. The fourth edition had just been printed and it immediately sold twelve hundred copies. Given that another edition would be necessary quite soon, Macmillan argued that "I think it very likely that a good preface at this point would be judicious. So please send me what you have written."[22]

As Seeley began writing the new preface, there were two prevalent, critical views of the book which corresponded to two assumptions about the author's theological identity. The first was that the author was well on his way to unbelief and was carefully concealing his true thoughts on Christianity in order to dupe the unsuspecting reader. And the second was the opposite formulation, that the author was concealing what were fairly orthodox Christian beliefs behind a rhetoric of free inquiry and historical criticism.

Seeley therefore found himself having to conduct a delicate operation in attempting to answer critics in his new preface. On the one hand, he clearly wanted to show that he had in fact thought a great deal about the Gospels from a critical perspective, that he engaged in a process of comparative analysis in order to evaluate the source base for his study of Christ. But on the other, in explaining his own critical analysis of biblical sources, Seeley had to be careful not to present himself as the radical freethinker that some of the book's more reactionary readers, such as *John Bull*, the *Quarterly Review*, and Lord Shaftesbury, were portraying. And, perhaps most problematic, in navigating these two positions he also needed to avoid alienating those orthodox readers who were praising *Ecce Homo*. This was a point Macmillan stressed after reading a draft of the preface: "remember all the warm friends you have among [the orthodox such as] Gladstone, the Guardian and the like, and where you can be explicit with a safe conscience it is better to be so."[23] Specifically, Macmillan was asking Seeley to be explicit about the historical reality of miracles. Seeley's statement that they should be regarded as "provisionally real" was clearly not convincing anyone, and Macmillan thought it would be best, in light of his new, "warm friends," to make a clearer statement in favour of their existence. Macmillan even suggested that, in reference to Fitzjames Stephen in *Fraser's*, that Seeley "might, *in the interest of Science*, claim that miracles should not be determined impossible."[24] This Seeley did, but within the context of further elucidating his methodology, which indicated a much more critical engagement with his sources than many of his freethinking critics might have realized.

After re-emphasizing the fragmentary nature of *Ecce Homo*, and the failure of some readers to recognize that the language was directed largely at a freethinking audience, Seeley addressed the fact that the book "contains no criticism of documents." But that did not mean, according to Seeley, "that the Author wrote without any criterion in his mind by which to test the veracity of the narratives from which he drew his conclusions, and that he simply assumed the truth of everything which struck his fancy or suited a preconceived theory."[25] There was, in fact, a method that guided the author's evaluation of the facts of Jesus's life: a comparative analysis of the four Gospels. "Out of these four writers," Seeley explained, "he desired, not to extract a life of Christ, not to find out all that can be known about him, but to form such a rudimentary conception of his general character and objects as it may be possible to form while the vexed critical questions remain in abeyance." Seeley stressed, moreover, the fact that there are discrepancies between the Gospels, which therefore "gives weight to their agreements." This was particularly true given the "wide divergence in tone and subject-matter of the Fourth Gospel from the other three," meaning that he felt that he could assume the truth of the statements where all four coincided with one another.[26] Seeley's main principle in evaluating the factual accuracies of the Gospels was to rely extensively on those statements on which all four agree. He then listed a series of twenty-one propositions that are "equally deducible, with scarcely the alteration of a word, from each of the other three Gospels."[27] These propositions largely include details about Jesus as a Messiah and as a worker of miracles.

In attempting to answer his rationalist critics such as Sidgwick, Seeley offered the possibility that these propositions may have been prejudiced by Christ's followers, who were attempting to establish Christianity and therefore set out to invent Jesus as Messiah and miracle worker. This notion was to be rejected, Seeley argued, as it "cannot reasonably be doubted" that "Christ did himself claim Messiahship." How else could his death be explained? Moreover, Seeley continued, it could not be doubted that it was believed "in his lifetime, and not merely after his death," that Christ was a worker of miracles. Perhaps with his warm friends in mind, Seeley went further still: "The fact that Christ appeared as a worker of miracles is the best attested fact in his whole biography, both by the absolute unanimity of all the witnesses ... and by countless ... special confirmations of circumstances not likely to be invented."[28] Seeley was insistent, therefore, that Christ's claims of being a Messiah and miracle worker were attested by a comparison of the Gospels and that such a comparison has provided "an irrefragable outline of that part of Christ's life which is discussed in [*Ecce Homo*'s] pages. The writer has adopted it as his framework, and has not attempted to add to it anything fundamental, but has simply sought to find in the Gospels matter illustrative of it."[29]

After having clarified the critical method of the book, Seeley went on to describe what he believed to be its very clear message, that the book was ultimately describing "a moralist speaking with authority and perpetuating his doctrine by means of a society." It was, therefore, "the union of morals and politics" that Seeley set out to show was ultimately "characteristic of Christianity." It is precisely this combination, according to Seeley, that distinguishes Christianity from other moral teachings such as that offered by the Stoics. "[W]ithout a society, and an authority of some kind, morality remains speculative and useless."[30] Seeley was not suggesting that Christ invented morals but asserting that "Christ has greatly elevated the generally accepted and, as it were, the attainable standard of virtue, and further, that he has set in motion a machinery by which, properly used, this standard may be elevated still further."[31]

Seeley concluded by pointing out, however, that the Church has not lived up to the promise of its original intentions. It has not failed. But it has not succeeded either.

> Since the Reformation it has acted rather as a dividing than a uniting influence, and further, that through a great part of its history it has been a too

> consistent enemy of freedom. It has been over and over again the main support of tyranny; over and over again it has consecrated misgovernment, and retarded political and social progress; repeatedly it has suppressed truth, and entered into conspiracy with error and imposture; and at the present day it fails most in that which its Founder valued most, originality; it falls into that vice which he most earnestly denounced, insipidity.[32]

With all this duly admitted, Seeley wanted to assure his readers that this did not mean "for a moment that the world can do without Christ and his Church."[33] *Ecce Homo* was written to show just how necessary was this view of Christianity so that the Church could be reformed in order to strengthen "the strongest and most sacred tie that binds men to each other."[34] Excluding the contentious issues surrounding the theology of Christianity was necessary in this regard, in order to "draw the attention of the public to that part of Christianity, and for a time to that part alone, in which almost all men are able on the whole to agree, and much of which the greater number of Christian teachers, by taking for granted, practically suppress."[35] This new preface was published as part of the fifth edition in June, though it should be stressed that, aside from the new preface, the main text was not altered in any way.

Was the new preface successful? It certainly clarified Seeley's own method for determining the facts of Christ's life. The *Spectator*, however, used the opportunity of a new preface to reinforce its own central criticism, arguing that "the author repeats, and even emphasizes, the radical defect of the whole book, which is the attempt to separate a purely spiritual legislation from the spiritual root of the legislation."[36] Meanwhile, it is unlikely that freethinkers or outright unbelievers found the clarified methodology a convincing one for establishing a truly historical foundation for the Christian Church along the lines envisioned by Jesus. While Sidgwick gave Seeley much direction in writing the new preface, in the end he was not convinced that it could adequately counter his criticisms. But if the book was not somehow offering a radical and factually accurate portrayal of the origins of Christianity, many were convinced that *Ecce Homo* did offer something different from what was currently on offer by the current factions of the Anglican Church. Moreover, it was becoming more difficult to imagine that the author fitted neatly into one of those factions.

Just as the fifth edition, complete with the new preface, was printed, John Henry Newman offered his own take on the debate about the book as well as the theological identity of the author. Newman was, it is safe to say, a divisive figure in British religious life, but he was also incredibly articulate and influential. In the 1830s and early 1840s he was the leading member of the Tractarian Movement, penning many of the works in the Tracts for the Times series that sought to re-establish older Catholic traditions in the Anglican Church. In 1845 Newman converted to Roman Catholicism, which confirmed for many liberal churchmen the papal direction that Tractarianism necessarily headed. It should be noted, however, that while some Tractarians followed Newman's example by converting to Catholicism, most others like Gladstone and Dean Church stayed within the Anglican fold. By the time Newman wrote his review of *Ecce Homo* in 1866, he had become the leading voice for an English Catholicism.

In the review, Newman pointed out that everyone was seemingly obsessed with discovering whether the author was "an orthodox believer on his road to liberalism, or a liberal on his road to orthodoxy."[37] Much of the confusion resulted from the fact that so many Tractarians (or members of the "Oxford school," as Newman called them) embraced Seeley's project, thereby being seemingly misaligned with Broad Churchmen against their usual allies in the High Church, who, as we have seen, came out in full force against *Ecce Homo*. But, according to Newman, *Ecce Homo* appeared at a fortuitous time for members of the Oxford school, "after the wearisome doubt and disquiet of the last ten years; for it has opened the prospect of a successful issue of inquiries in an all-important province of thought, where there seemed to be no thoroughfare."[38] Moreover, Newman continued, there was more fluidity between the great majority of the membership of the Tractarian and Broad Church parties of the Church than had been typically recognized. The reception of *Ecce Homo* helped bring that to light.

> Distinct as are the liberal and catholicising parties in the Anglican Church both in their principles and their policy, it must not be supposed that they are as distinct in the members that compose them. No line of demarcation can be drawn between the one collection of men and the other, in fact; for no two minds are altogether alone, and, individually, Anglicans have each his own shade of opinion, and belong partly to this school, partly to that. Or rather, there is a large body of men who are neither the one nor the other … they range from those who are almost Catholic to those who are almost Liberals. They are not Liberals, because they do not glory in a state of doubt; they cannot profess to be "Anglo-Catholics," because they are

> not prepared to give an internal assent to all that is put forth by the Church as truth of revelation. These are the men who, if they could, would unite old ideas with new; who cannot give up tradition, yet are loth to shut the door to progress; who look for a more exact adjustment of faith with reason than has hitherto been attained; who love the conclusions of Catholic theology better than the proofs, and the methods of modern thought better than its results; and who, in the present wide unsettlement of religious opinion, believe indeed, or wish to believe, Scripture and orthodox doctrine, taken as a whole, and cannot get themselves to avow any deliberate dissent from any part of either, but still, not knowing how to defend their belief with logical exactness, or at least feeling that there are large unsatisfied objections lying against parts of it, or having misgivings lest there should be such, acquiesce in what is called a practical belief, that is, believe in revealed truths, only because belief in them is the safest course, because they are probable, and because belief in consequence is a duty, not as if they felt absolutely certain, though they will not allow themselves to be actually in doubt.[39]

It was to this group of men that *Ecce Homo* appealed. Newman argued that to this large class *Ecce Homo* "comes as a friend in need." The book therefore had the merits of promising to provide just what this large group of Christians needed in order to rejuvenate their belief, and Newman believed that the unknown author deserved heaps of praise for attempting this difficult feat. And ultimately, Newman argued that the book would "subserve the cause of revealed religion."[40]

What Newman clearly liked about the book was the fact that Seeley had acknowledged "the visible Church as our Lord's own creation, as the direct fruit of his teaching, and the destined instrument of his purposes."[41] This, of course, was precisely what some freethinkers like Sidgwick did not appreciate about *Ecce Homo* because such an argument was based on the assumption of teleology, that Jesus knew all along that he was to found a church in his name and that it would become an instrument for his teaching.[42] No doubt this was one of the reasons why some members of the Oxford school found *Ecce Homo* useful because it did not reject out of hand some of their own Catholic-oriented beliefs. What Newman did not appreciate, however, was the fact that this was a preliminary study, which of course was one of the main defences Seeley relied upon to justify the book's seemingly ambiguous conclusions. But for Newman, its preliminary status did not justify the "patent theological errors," the "bad logic," the "rash and gratuitous assumptions," or the "half-digested thought" that seemed to fill its pages. The author would have served his purposes much better, and avoided much criticism, had he spent more time digesting his research and exploring his own beliefs about that research before publishing. "[W]e are obliged to conclude that it would have been much wiser in him if, instead of publishing what he seems to confess, or rather to proclaim, to be the jottings of his first researches upon sacred territory, he had waited till he had carefully traversed and surveyed and mapped the whole of it."[43]

About the author's identity Newman argued that he was, because of many of his rushed assertions and perplexing arguments, an "enigma." For Newman, however, what was at issue was whether the author "comes to us as an investigator or a prophet, as one unequal or superior to the art of reasoning."[44] Repeating a point made in the *Guardian* review, Newman admitted that "[t]here is always a danger of misconceiving an author who has no antecedents by which we may measure him." Given that he remained in the shadows, it was necessary to take the author completely at his word. In that regard, however, "we can but wish that he had kept his imagination under control; and that he had more of the hard head of a lawyer, and the patience of a philosophy. He writes like a man who cannot keep from telling the world his first thoughts, especially if they are clever or graceful; he has come for the first time upon a strange world, and his remarks upon it are too obvious to be called original, and too crude to deserve the name of freshness."[45]

In the final analysis for Newman, however, the book was clearly written by a Protestant and ultimately could only be about the kind of concerns that typically motivate Protestants about their belief. It is to be expected that Newman's review was ultimately a foil to present Catholicism as offering the key solutions to the theological conundrums engendered by *Ecce Homo*. "[W]e will confess that Catholics," argued Newman, "kindly as they may wish to feel towards him, are scarcely even able, from their very position, to give his work the enthusiastic reception which it has received from some other critics. The reason is plain; those alone can speak of it from a full heart, who feel a need, and recognise in it a supply of that need. We are not in the number of such; for they who have found, have no need to seek … Catholics are both deeper and shallower than Protestants; but in neither case have they any call for a treatise such as this *Ecce Homo*."[46]

Despite this criticism, Macmillan was quite pleased with Newman's review. "Newman's article is very interesting," Macmillan wrote to Seeley. "You have hardly had any higher compliment than his writing at such length about you."[47] It was certainly not a scathing attack. And Newman made some useful observations with regard to what he perceived to be the fluid boundaries between competing schools of Anglican parties, though this no doubt was also self-serving, as it implicitly justified his conversion to Catholicism. But while Newman made it clear that the book was not for Catholics, not unlike the way that Sidgwick and Stephen made it clear that it was not for freethinkers, he pointed out that what it did offer was something different for those Protestant believers who did not sit so clearly with one particular camp or another. This was ultimately an attractive perspective put forward by an unlikely source. However, it was a difficult perspective to establish when most interpretations of the book were so clearly determined by speculations about the author's identity. It is to such speculation that we will now turn.

Chapter Eight

Shrewd Conjecture

The most likely and the most unlikely persons of various professions, positions, and creeds are named or guessed at as the writers whose brains and pens the work is said to have emanated which of late has stirred the minds of all who speak the English language at home or abroad.

– "Ecce Homo," *Melbourne Punch*, 30 August 1866

By the summer of 1866, *Ecce Homo* had by all accounts become a remarkable literary sensation that rivalled a handful of much better remembered books published during the mid-Victorian period. An oft-made comparison at the time was with the *Vestiges of the Natural History of Creation* (1844), a book about an equally contentious subject (evolution) that was written with much the same purpose, to bring contemporary scientific and theological knowledge into better alignment. It was not the subject matter that engendered the comparisons, however, but rather the fact that both *Vestiges* and *Ecce Homo* were anonymously published, and that both engendered an enormous amount of interest about the identity of their masked authors. "No anonymous book, since the 'Vestiges of Creation' (now more than twenty years old)," argued Gladstone in his extended treatise on *Ecce Homo*, "indeed, it might almost be said no theological book, whether anonymous or of certified authorship – that has appeared within the same time interval, has attracted anything like the amount of notice and of criticism which have been bestowed upon the remarkable volume entitled 'Ecce Homo.'"[1]

As was also the case with *Vestiges*, much of the excitement surrounding the publication of *Ecce Homo* had to do with the ambiguous identity of the masked author, an identity that was shaped not only by the publisher and author but also by readers, who often projected an identity onto the text based on their own reading. So speculation about the author was, in one sense, serious business, as it attached a particular meaning to the text. But it could also be simply entertaining to think about and discuss plausible candidates with friends and colleagues over dinner or at reading clubs or in correspondence. Periodicals and newspapers attempted to cater to these kinds of conversations in their "literary gossip" columns, shedding light on the social side of the publishing world that readers were being invited to join by engaging in literary discussion that often went beyond the text itself. Speculating about anonymous and pseudonymous authorship was a central subject of this communal discourse.

Engendering what we might call extra-textual discourse about a given work was important for publishers looking to broaden the appeal of a book beyond the rather narrow readership of a particular genre. No doubt the subject of Seeley's book meant that it already had a large readership to draw from, but it became a literary sensation because its authorship became such a wide subject of conversation. By May, with *Ecce Homo* about six months old, the periodical and newspaper presses began making precisely this point. "Rumours are still flying about respecting the authorship of 'Ecce Homo,'" the *London Review* reported in its "Literary Gossip" section, "and there can be no doubt that the well-kept secret has had a great deal to do with the large sale of the work. We learn from various booksellers that, in nine cases out of ten, the second question asked by the intended purchaser is, 'Do you know who is the author?'"[2] Speculating about the answer to this question became a topic of conversation in the bookshops, a conversation that was then reported in the press, thereby creating further interest in the question and the book itself. The *Sheffield and Rotherham Independent* similarly reported in May that "The curious are still trying to piece the mystery which surrounds the authorship of 'Ecce Homo.' But the secret is well kept ... Whoever the author, the book has reached its eighth thousand, and can hardly be had at Mudie's, so great is the demand."[3]

Such bookshop gossip led to a wide range of suggested candidates, from the plausible to the downright absurd. "He is known to be an Oxford man, and believed not to be a clergyman," the *Sheffield and Rotherham Independent* continued. "Some trace in the pure and classic style of the book the hand of [James Anthony] Froude; others, most erroneously, have fathered it on Professor Goldwin Smith."[4] The *London Review* gave a brief history of the speculation, writing that first "Vice Chancellor Page Wood was eagerly chosen for the post; then came

Mr. George Waring, of Magdalen Hall. A later favourite was Professor Goldwin Smith, and his recent visit to America and sojourn with Emerson has been dwelt upon with considerable gusto, as throwing some light upon the authorship. The last favourite will strike many persons with surprise. It is no other than the Emperor Napoleon III. whom [*sic*] many persons in Paternoster Row roundly assert wrote the book in French, and then sanctioned its translation into English."[5] This latter suggestion was no doubt the most absurd, and perhaps because of that received a fair bit of press, with short notices of Napoleon III's suspected authorship appearing in the *Glasgow Herald*, the *Pall Mall Gazette*, the *Bradford Observer*, and the *Birmingham Daily Post*.[6] The proof seemed to be provided by the *Scottish Guardian*, which included a reprint of a paper supposedly written by Napoleon and first published in France in 1841, anticipating many of the arguments put forth in *Ecce Homo*.[7]

Certain suspected candidates were often reported as being the author, only to be refuted by the person in question, leading to the inevitable retractions. After reporting that George Waring was the author, *John Bull* was forced to write later that its report was now being denied by the *Oxford Times*.[8] "We have undoubted authority to contradict the report which has been set afloat to the effect that 'Ecce Homo' was written by Mr. George Waring, of Magdalen Hall. Mr. Waring is a member of the Universities of Oxford and Cambridge, and took a very 'good first' in Greats and the former University, but he has nothing whatever to do with the authorship of 'Ecce Homo.'"[9] Page Wood's name was also floated as a possibility, only to be quickly denied. The *Pall Mall Gazette* reported that "The *Solicitors' Journal* contradicts on authority a report that Vice-Chancellor Sir P. Wood is the author of 'Ecce Homo' we are all reading and wondering so much about just now."[10] Other reports stated that Wood was adamant that he was "not the author of that foolish book, 'Ecce Homo.'"[11] In a similar vein, the *Star* reported that the Arabic scholar and former Jesuit Gifford Palgrave was the author. The *Pall Mall Gazette* heaped scorn on this notion, however, suggesting that the several factual errors in the report, such as that Palgrave went to Trinity, Cambridge, and not Oriel, Oxford, made it entirely unreliable.[12]

The speculation as to the authorship was not confined to the English press, as the colonies engaged in their share of spreading the rumours about *Ecce Homo*'s authorship. The *Friend of India* (later to become the *Statesman*) reported that there were rumours the author was Henry Maine, "who will be astonished to find himself named by the Baptist paper, the *Freeman*, as the probable author of 'Ecce Homo.'" The *Friend of India* was sceptical of this suggestion, however, believing that "there is no doubt that a Curate of the Church of England is the author."[13] The news reached Australia as well: the *Australasian* reported that "'Ecce Homo' has reached its twelfth thousand, and still the author's name is a profound and tantalising secret. The public had paid £6,000 for the book, and it is probable that at least an equal amount will yet be received before the mystery is solved."[14] Meanwhile, it was left up to the *San Francisco News Letter* to discover that the author was the Unitarian minister John Hamilton Thom.[15]

The *Melbourne Punch* was also fascinated by the curiosity surrounding the "probable authorship of the book of the day, *Ecce Homo*." "The most likely and the most unlikely persons of various professions, positions, and creeds are named or guessed at as the writers whose brains and pens the work is said to have emanated which of late has stirred the minds of all who speak the English language at home or abroad." However, Mr Punch had uncovered "unmistakable evidence which proves that the book in question is of colonial authorship, a fact which the advocates of native industry will no doubt hail with delight." Mr Punch even included a long list of possible candidates, including "eminent statesmen, eloquent divines, philosophers, and men of letters," though he admitted that he could have come up with some more likely candidates "had he felt so inclined, but as it happens he doesn't."[16] The suggestion, of course, was that names were being guessed at without any real evidence to back up such speculations.

While this is certainly partly true, some of the speculation was a reflection of the meaning a given reader was projecting onto the text. Lord Shaftesbury, who we have seen was a highly vocal critic of the book, was convinced that the author was James Anthony Froude.[17] The choice of Froude is telling because the historian was a favoured target of orthodox churchmen in the mid-Victorian period. This was because his elder brother, Richard Hurrell Froude, had been one of the early leaders of the Oxford Movement, and was much admired for his saintly character before he died just as the movement was at its peak. The younger Froude had seemed to be following in his brother's footsteps, contributing in particular to Newman's series on the Lives of Saints, before seemingly turning against his family and his Tractarian friends by writing the controversial crisis-of-faith novel *Nemesis of Faith* (1849). About a Church of England cleric who could not reconcile his growing doubts because of the dogmatism of the Tractarians on one hand and the literalism of the evangelicals on the other, the book was famously burned at Exeter College Hall and led to the author's forced resignation of his fellowship. While Froude would himself move beyond this scandal and become the best-selling author of a twelve-volume *History of England*, this episode would continue to play a role in the reception of his works.[18] As Shaftesbury thought the book was vomited from the jaws of hell, who better than Froude to have written it?

In an altogether different vein, those orthodox churchmen who found much to admire in *Ecce Homo* were understandably inclined to assume that the author was a much more respectable figure. The Tractarian R.W. Church, for instance, thought that the author was John Henry Newman. Interestingly, Newman had been Church's mentor long before he had converted to Roman Catholicism. And Church still greatly respected him and his writings despite the fact that the two men were now in different Christian denominations. This mistaken assumption about the authorship of *Ecce Homo* no doubt played a role in Church's sympathetic reading of the book, a reading that was then expressed in several columns of space in the *Guardian*. Given the role that Church's review played in the reception of the book (see chapter 5), it is a compelling thought that it was written under the assumption that the author of *Ecce Homo* was Newman. Perhaps Church believed that this was Newman's attempt to show that the kernel of Christianity, namely Christ's character, could unite both Protestants and Catholics. Such a view seems incredible in hindsight, particularly when considered against Newman's much later review of the book, which said that *Ecce Homo* simply did not speak to Catholic concerns. Moreover, W.R. Nicoll writes that this mistaken assumption on Church's part was particularly difficult to understand, given that "no one understood Newman and all the secrets of his style and thought so well as Church; and yet he fell into this strange blunder."[19] Church, however, was not alone in thinking that Newman wrote the book. Macmillan expressed surprise at this fact, writing to Seeley that "[i]t is odd how many people have thought of Newman."[20] But it is less odd when we consider that many of the guesses about the author's identity were reasoned assumptions based on particular readings of the book.

George Eliot was also a favoured candidate. This assumption was likely based on the fact that she had translated Strauss into English and was also an early supporter of positivism. Those believing that *Ecce Homo* combined the critical approach of Strauss with the positivist epistemology and humanism of Comte would have been hard pressed to find a more suitable candidate. Something about this rumour did not seem quite right to Matthew Arnold, however. After a conversation with James Martineau, he was convinced that it could not be Eliot.[21] Many others likely made the same realization after Sidgwick's review in the *Westminster* confirmed that *Ecce Homo* did not follow an adequate critical method.

By June, however, as the identity of the author had yet to be disclosed, some of the speculation took on a more investigative nature. This was particularly true surrounding the rumour that the author was the editor of the *Spectator*, R.H. Hutton, the same Hutton who wrote the first review of *Ecce Homo* that caused so much trouble for Macmillan and Seeley (see chapter 5). This "shrewd conjecture" was made after some seemingly very fine deductive work that led to Hutton as the only possible candidate. "We believe," the *Bookseller* reported, "that we shall not be very far from the mark when we guess that he [the author of *Ecce Homo*] will probably be found in the editorial chair of a London newspaper, and that he formerly edited a review which we regret to say is now discontinued. In early life the gentleman in question was a Unitarian, closely connected with a celebrated literary family of that denomination; later in life his views became more advanced, while his faith contracted; but more recently he has attached himself to the Church of England, and will be frequently seen attending the ministry of the Rev. F.D. Maurice. If this guess proves correct, many of our readers will have no difficulty in recognising the writing of 'Ecce Homo' by the above description."[22] A similar deduction made by the *Patriot* on 19 July led to an equally interesting suggestion: "The shrewdest guess yet made at the authorship of this remarkable book [*Ecce Homo*] is that which ascribes it to Professor Seeley, who fills the Latin chair in the London University College."[23] Luckily for Seeley, this conjecture was largely ignored. His secret was still safe, for the time being.

Just how the authorship managed to be kept a secret also became a subject of wild speculation. It astonished many observers that such a sensational book could hide its authorship so thoroughly when so many readers seemed set on discovering the secret. It was widely believed that, according to the *London Quarterly Review*, "even the publishers have not been entrusted with the name of its author," and this apparently helped explain why the secret of *Ecce Homo*'s authorship lasted into the summer.[24] Writing from the Athenaeum Club on 14 May, the London correspondent for *Trewman's Exeter Flying Post* claimed that the "Ecce Homo controversy still continues." Sir Page Wood disclaims the "silly book" while Lord Shaftesbury says that it was "vomited from the jaws of hell." But the really interesting question for the correspondent was this one: "who is the author?" Apparently, "[v]arious traps have been laid to catch him, but he has escaped them all. Even Macmillan, the publisher of the book, does not know him. He says the MS. is in a woman's hand, that it came to him through Dean Stanley, to whom all the proofs were sent, and through whom all the correspondence passes, and that the Dean will say no more than that the author is a curate."[25] This latter story, about the MS coming to Macmillan through Stanley in a woman's hand, gained legs and was also later reported on in the *Friend of India*.[26]

By the late spring and early summer, this sort of explicit speculation surrounding the authorship of *Ecce Homo* was making its way into the more serious periodical review literature. The first to make this speculation central to the reception of *Ecce Homo* was A.P. Stanley, who as we have seen was one of the main figures associated with the publishing of the book. Writing under his initials "A.P.S." in *Macmillan's Magazine*, Stanley pointed out that an incredibly diverse list of individuals had been attributed with the authorship, including "the most celebrated of Roman Catholic divines, the most learned of Roman Catholic laymen, we know not how many Nonconformist ministers, three Essayists and Reviewers, an Archbishop of York, innumerable young fellows of Colleges, a Republican Professor, a female novelist, a leading journalist, an Irish novelist, a Scottish poet, a Scottish duke, a Master of Trinity, a Dean of Westminster, an Attorney-General, a Poet Laureate, a Chancellor of the Exchequer, a High Church Vice-Chancellor, a law stationer, a chemist, an unknown sea captain, and the Emperor of the French."[27] For Stanley, the great diversity that was to be found in this growing list of possible authors suggested something about the meaning of the book rather than about its secret authorship. It suggested that *Ecce Homo*'s message was one that clearly spoke to a wide range of Christian denominations and parties. This was why, for Stanley, it was so important that the book was published anonymously and that it should remain so. "In our day," he continued, "when theological polemics attach so much more to names and stations than to doctrines, when respect of persons is so much more powerful than respect of truth, it is doubly important that, at least until the work before us is completed, the vultures of party controversy should be kept at bay."[28] Without a name to attach to the book, Stanley argued, readers were given more freedom to think about it as they wished. But should that name become known, he worried that the debate would become one between the familiar Church parties while the novel perspective offered by the book would be all but ignored.

It should not be surprising that Stanley was here parroting the "official" line for maintaining anonymity, that the author did not want his book to be associated with a particular name that would then bias its reception. This was an argument that was originally deployed by Macmillan in response to Gladstone's much-prized Christmas-day letter. And now it was an argument being explicitly connected to the diverse speculation about the unknown author. The speculation itself was now evidence for the wide-ranging appeal of Seeley's message, a view being expressed in *Ecce Homo*'s publisher's house magazine. This was no doubt in part why Macmillan hoped that Stanley would publish his review in the *Edinburgh Review*,[29] not just because it would reach a wider audience but also because it would then be less directly connected to the publisher of the book. This is not to say that Stanley's views about *Ecce Homo*'s anonymous status were not genuine; to the contrary. Indeed, he told Macmillan upon first reading the book that he hoped the author's name "should not be disclosed till after the appearance of the 2d volume. When once it is known, the genuine interest that has been felt will be diverted ... [into] party classes – and the effect of the book will be partly marred."[30] But this was certainly a view that both Seeley and Macmillan found attractive for the continued sales and reception of the book. Moreover, Stanley's review was widely reported in the newspaper press, in large part because now the explicit speculations about particular individuals were being connected to interpretations about the larger meaning of the book.[31]

Other reviews around this time followed Stanley's lead by essentially acting as commentaries on the reception of the book in connection with the speculation about its unknown authorship. The *Fortnightly Review*, for instance, summed up these views well by suggesting that *Ecce Homo* "has been called an infidel book artfully constructed to lull the suspicions of the orthodox; it has been called an orthodox book, disguised under a veil of free inquiry, to attract the notice of skeptics." Neither of these theories was appropriate, argued the reviewer, who found them "insolent, paradoxical, and mutually confuting."[32] What the previous reviewers had failed to appreciate was that it was clearly written by someone who did not fit into these pre-existing positions. "It is an outside book, written by one who found himself cast free from all the religious moorings of his youth, and who determined to start on a voyage of discovery, instead of having recourse to the regular theological guides."[33] The author, therefore, offers a middle ground between the extremes: he does not want to destroy but rather to build. And yet he also wants to lay a foundation for the Church of the future not for the Church of the past. "[H]e is a spiritualist, yet ardently scientific; an idealist, yet always planting his foot on reality and fact."[34]

In a pamphlet on *"Ecce Homo" and Its Detractors*, George Warington found it similarly amusing how readers sought to search out the various clues and hints in the text that must surely indicate the author's theology. "And very eager have many inquisitive readers been to gather up these hints, and form conjectures as to who the author may therefore probably be. The wonderful variety of the conclusions thus arrived at is sufficient proof how well the author has in fact kept clear of debatable theology. He has been reputed to be High, Low, Broad, and what not else." Warington's own conclusion was much like the *Fornightly*'s, that the author belongs to none of these schools of thought, but that "he is one of those, who, having been brought through doubt to a more living apprehension of the realities of truth than ordinary, is raised above parties altogether, and can afford to look down upon and do without the pet phrases and conventional forms which these have coined to represent their favourite and characteristic views."[35] What *Ecce Homo* was offering was something so entirely new that the reader was simply failing to comprehend it, thereby relying on the old divisions to make sense of it and its author.

This was a message that was even more appealing to *Ecce Homo*'s readers in the United States, where it began to be published in April of 1866. To advertise the fourth "American edition," *Ecce Homo*'s Boston publisher, the Roberts Brothers, issued a promotional pamphlet that referred to it as "emphatically, the sensation of the day in literature. Published anonymously in England, it has already run through seven editions there, and is still the subject of universal comment." The pamphlet also described the quick sale of three previous American editions, which had been "almost entirely confined to Boston and the immediate vicinity." What was driving interest in the book, of course, was the "excitement ... to ascertain who the anonymous writer is." The pamphlet continued that "One of the most remarkable features of this book is the entire absence of any expression or opinion, which would indicate to what religious denomination the author belonged. It is impossible to locate him in theology, and he does not allude to an author or a sect in Christendom. Hence it is read by believers in all creeds with the same interest."[36] In the following edition, the Roberts Brothers were a bit more specific about how *Ecce Homo*'s readers were "prophesying him [the author] to be either Trinitarian, Unitarian, Romanist, Quaker, or Infidel."[37]

Of course, suggesting that readers did not know how to categorize the author along a particular theological trajectory did not necessarily mean that the author was somehow transcending traditional party allegiances. Indeed, it could suggest instead that the author himself was simply confused about his own views and unable to present them in a manner that was theologically coherent. This was the opinion of the *Christian Remembrancer*, which presented its own theory about the author's identity. It was no doubt true, argued the reviewer, that the unknown author was not a radical Arnoldian, as had been suggested by one section of the media, nor was he a High Church Anglican who had wrapped himself in the garb of a more radical persona, as was argued by another set. Based on the evidence that the author himself presents in the book, he is most likely "a layman – a man of high, and most probably academical, education. He is one who has read and thought deeply and earnestly upon a large number of social, philosophic, and religious problems. We should not, however, gather from his work that he has dived very profoundly into theological lore. But he has much power of combination, and no small portion of originality. The arrangement, the conciseness, the form of expression, all seem to point to the pen of the practiced writer." The problem, however, was that the book was written in a particular way to produce an effect upon the reader about the purity of the author's convictions. The reviewer argued that the book was specifically "calculated to leave upon the mind of an unbiased critic the conviction that he has been brought into contact with a spirit of remarkable purity and honesty."[38] The author, therefore, has not sought to deceive anyone about his own doubts and beliefs. But he has, perhaps, deceived himself about what those beliefs may be. This is no better evidenced than by the fact that "among all his numerous and gifted critics, no one is able positively to say what this definite opinion is. To one he seems to hold views concerning the world's Redeemer not higher than those of Marcion; to another he appears to symbolise with Dr. Channing; to a third to be leading his readers onward through the contemplation of the humanity to the divinity of Christ our Lord. And each reviewer, in his turn, is able to make out a plausible case on his behalf of his own theory." The contradictory theories about the author's identity and argument, therefore, are not evidence that he had found some new way to transcend old divisions but more obviously that his argument is simply incoherent.

From the perspective of the *Christian Remembrancer*'s reviewer, the self-assuredness of the style of *Ecce Homo* led readers to believe that the author in fact had a clear understanding of his own views, such that it enabled otherwise careful reviewers to unknowingly project their own theories into the empty space left by the author's ambiguity. There was no doubt in the reviewer's mind, however, that *Ecce Homo* was "unquestionably ... the book of the season. This is shown not merely by the extraordinary sale of the work, but by the number of very distinguished persons who have undertaken to review it – a list which is commonly reported to comprise an eminent Fellow of one of the first colleges in Oxford, the Bishop of a western see, Dr. Newman and the Dean of Westminster. Mr. Vaughan and Mr. P. Bayne must be added to the series; as also one of the few Presbyterians who has a real claim to be considered as a theologian, Principal Fairbairn of Glasgow." The reviewer doubted, however, that this celebrity would endure. It was certainly rare that theological treatises enjoyed such popularity to begin with. But given that "the writer's aim is, after all, so exceedingly uncertain," the reviewer doubted that the popularity of the book would last much longer.[39] "When the first dazzling effects of the author's brilliancy have passed away, we believe that few will be found to consider the 'Ecce Homo' a complete success. But far more intense, we venture to predict, will be the failure of one who shall attempt to set forth Christ 'as the creator of modern theology and religion,' without announcing clearly and dogmatically what he believes that Founder to be."[40]

The *Christian Remembrancer* was no doubt right that *Ecce Homo*'s celebrity could not endure for much longer, particularly given the way in which its popularity was largely driven by the speculations about who the author might be. What would happen to the book once the author's name became known? Would the debate about its meaning, as Stanley believed, suddenly take the shape of traditional Anglican disputes? Or would the debate surrounding the book simply disappear, as was suggested by the *Christian Remembrancer*? Interestingly, despite the fact that there was much surprise expressed about how long the secret authorship had been maintained, no one was asking *whether* the authorship would eventually be discovered. The question being asked was always *when*. Indeed, the book was being marketed as if there would eventually be some grand unveiling of the author's identity, perhaps, as Stanley suggested, with the publication of the promised second volume that would deal more directly with the contentious issue of Christian theology that had been ignored in *Ecce Homo*. Unfortunately, the discovery of the author's identity proved to be much less deliberate, even if it should have been easily foreseen as the inevitable result of Macmillan's risky marketing strategy.

Chapter Nine

White Lies

Every man is conscious that of the morality which he theoretically holds there is one part which he always and easily practices, and another part which he often neglects.

– "Preface to the Fifth Edition," *Ecce Homo* (1866)

Given how useful the speculation surrounding *Ecce Homo*'s authorship was for promoting the book, exploiting that speculation became a central plank of Macmillan's marketing strategy. That strategy seemed to be simply based on the assumption that getting people thinking and talking about who the author was increased the interest, and hence the sales, of the book. While this strategy worked very well for a time, it also necessarily put the publisher and author in uncomfortable situations, particularly when Seeley began to be posited as the author. In this regard, the marketing of *Ecce Homo* brought to light one of the more disreputable aspects of Victorian publishing, namely the deception and even outright lying that seemed to result from producing an anonymous book. These were, moreover, not the kind of associations that were beneficial for a book that was essentially an extended meditation on Jesus Christ's moral philosophy. But clearly these were risks that Macmillan was willing to accept in order to take full advantage of *Ecce Homo*'s anonymous status. And, at least initially, Seeley was as well.

One particularly stunning example of Macmillan's marketing strategy at work was actually publicized in Stanley's review of *Ecce Homo*. "There is a legend floating around London," Stanley argued, "that the publisher invited sixteen persons to dinner, to meet the author of 'Ecce Homo,' who returned home no wiser than they came."[1] And this was despite the fact that Seeley was one of the sixteen present. Macmillan's biographer has confirmed this story, arguing that Macmillan "kept his word, for Seeley was present; but he kept his counsel also, and his guests, having vainly searched one another's faces for the marks of authorship, went out, none the wiser."[2] While Macmillan's biographer has argued that this was clear evidence of Macmillan's ability to "guard" his author's wish for privacy, it seems more clearly to have put it in jeopardy for the sake of generating interest in the book.[3]

For the most part, however, Macmillan's strategy seemed to involve simply getting interested readers to speculate about who the author might be as a way of creating a further layer of intrigue about the subject of the book. He wrote to historian Edward Freeman, for instance, who was in the midst of editing a series of books for Macmillan, and, unprompted, asked him who he thought the author might be. "I don't know who wrote 'Ecce Homo,'" Freeman maintained, "nor have I seen the book," but he had heard from Lord Shaftesbury that it was "the father of lies," meaning his nemesis James Anthony Froude.[4] Macmillan spoke about the unknown author with Matthew Arnold, who came away from their conversation convinced that it must be a Cambridge man, though he settled not on Seeley but on William Hepworth Thompson, the master of Trinity, Cambridge.[5] Macmillan also discussed the possible author with Tom Hughes, who happened to be the anonymous author of *Tom Brown's School Days* (1857), which Macmillan also published. Hughes apparently was led to believe that the author was James Campbell Shairp, a master at the Rugby school for boys.

These mistaken guesses were typically not rejected by Macmillan (though some were, as we will see below) but rather used as opportunities to further circulate possible names that Macmillan could then claim others had suggested. This was also a way for Macmillan to keep up the ruse that he did not know the author, while increasing interest in the speculation. This strategy reached absurd heights when Macmillan decided to confront one of the supposed authors of *Ecce Homo*. He relayed this episode to Seeley on 13 February 1866: "Among the conjectures as to the authorship of 'Ecce Homo' was one which Tom Hughes made, namely that it was an old Rugby master of his Shairp by name … I wrote [Shairp] a week ago asking if it was true that he was the author, as tho E[cce] H[omo] had come to me anonymous and other chaff of that sort."[6] Shairp responded emphatically: "No! My Dear Sir. I did not write Ecce Homo!" (See figure 9.1.) He was, however, not offended by the suggestion and went on to guess about who the author might be, engendering just the speculation Macmillan was seeking. "Whoever the great unknown

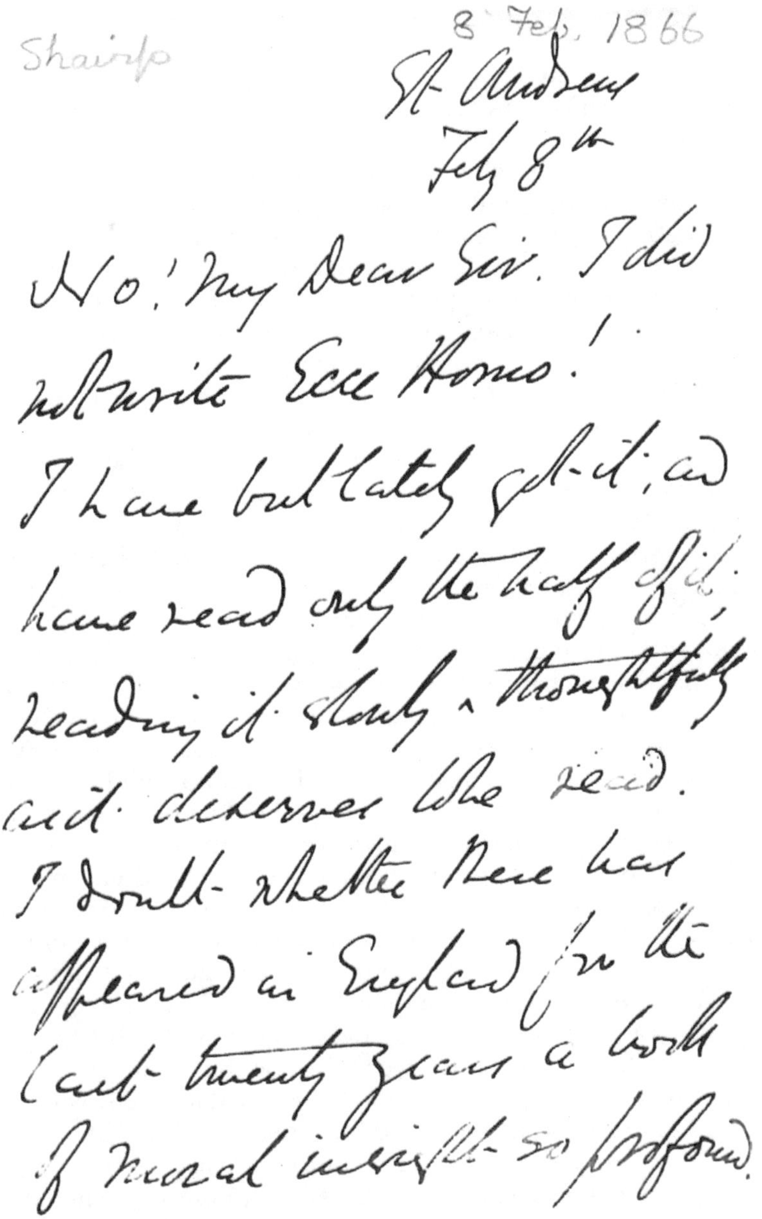

Shairp

8 Feb. 1866

St Andrews
Feby 8th

No! My Dear Sir. I did not write Ecce Homo!
I have but lately got it; and have read only the half of it; reading it slowly & thoughtfully and it deserves to be read.
I doubt whether there has appeared in England for the last twenty years a book of moral insight so profound.

Figure 9.1 James Campbell Shairp to Macmillan, 8 February 1866, Seeley Papers, MS 903/3A/1. Courtesy of the Senate House Library, University of London.

is," Shairp wrote, "he must be a man of rare purity, depth, [etc] … I am quite sure it is not Dr. J.H. Newman … When you have detected the great unknown, if it is not to be kept a secret, I shall hope to hear his name."[7]

While Shairp's exchange with Macmillan has led to a great deal of confusion in the secondary literature about Macmillan's knowledge of *Ecce Homo*'s authorship,[8] the response Macmillan received from Shairp was just what he was hoping for. Of course he knew that Shairp would deny writing the book. But his purpose in writing Shairp was not to gain some sort of false admission. His purpose was to get Shairp interested in the question of *Ecce Homo*'s authorship. Shairp went on to say that Macmillan had "stimulated my curiosity, great before, to a greater pitch."[9] For Macmillan, increasing the speculation about the authorship could only help generate interest in the book itself and thereby increase its sales. As he wrote to Seeley about further inquiries about his authorship, keeping up the "uncertainty … will do quite well for our purpose."[10] But Macmillan's purpose, which was to sell copies of the book, did not always overlap with Seeley's, which was – first and foremost – to avoid being discovered as the author.

This would become apparent almost immediately after the book was published. Macmillan claimed to one correspondent that in order for the authorship to be kept a secret, "as a rule," he would not say "no, or yes, to any one."[11] In other words, Macmillan claimed that he would simply let speculation flourish about any and all authors without giving any hint about who the author might or might not be. This surely seemed sensible enough at the time. However, from the start, one of the favoured names to be suggested as author was that of Goldwin Smith, the cranky Broad Church Oxford professor of modern history. On 13 December 1865, with *Ecce Homo* on the bookshelves for barely a month, Macmillan reported to Seeley that Dean Stanley "guesses Goldwin Smith. I have told him his guess is wrong."[12] He explained his rationale for doing this ten days later when he told Seeley that Smith was not just favoured by Stanley but was assumed to be the author by readers in Glasgow, Cambridge, and Oxford. The fact that Smith was so widely believed to be the author placed Macmillan in a bind. Not only was Smith known to harbour views similar to those of *Ecce Homo*, but also the writing styles of Seeley and Smith certainly bore some similarities. Many readers were therefore almost convinced that Smith must be the author. Should the speculation centre on one name, it would likely stop, and Macmillan, of course, wanted it to continue. But perhaps more problematic was the fact that Smith was published by Macmillan.

Macmillan therefore felt obligated to discount any false speculation that might be directed at one of his authors, to "disabus[e] every body's mind on" the possibility of Smith's authorship. "I am doing so because there is so much intimate connection between him and me that people would conclude by my silence that he wrote the book. He would be quite decided on and curiosity would cease. Besides, there is a certain superficial resemblance in your writing styles."[13]

Denying certain candidates outright not only undermined what was supposed to be a general policy that Macmillan was following in order to protect Seeley's name, but it also contradicted the argument that Macmillan at times suggested – that he himself did not know who the author was. Moreover, this was an argument that Macmillan only half-heartedly sought to maintain. This is most apparent in a letter Macmillan received from Rose Kingsley, Charles Kingsley's eldest daughter. She wrote to Macmillan inquiring precisely about the conflicting reports she was hearing about what Macmillan may or may not have known about the author's identity. "There are so many reports going on about … the Authorship of Ecce Homo," Rose wrote to Macmillan, "and some people go so far as to say that even you don't know who it is by, that I determined to ask you. Not that I want to know who the Author is if that is a secret but I want quickly to hear from you that you do know who it is. Because I thought you said you knew at Eversley."[14]

If this episode points towards a flaw in Macmillan's strategy to engender speculation about the authorship, there was an even greater one that seemed to catch both Macmillan and Seeley off guard, namely how to handle the dilemma of denying a correct guess. This did not become an issue for several months, as speculation tended initially to centre on well-known and divisive figures such as Newman or Smith or Froude. But as time went on, many of these figures were discounted. Obviously there were those like Smith, whose name Macmillan immediately discouraged as a possible author. Others were removed from the list in a more indirect fashion: they reviewed the book themselves. Indeed, it was not until Newman reviewed *Ecce Homo* in the *Month* in June that speculation about his authorship ceased. And once speculation turned to second- and third-tier public figures, denials of a more public nature ensued, such as the case of George Waring denying his authorship in the *Oxford Times* (see chapter 8).[15] Seeley was acutely aware of these public denials because they meant that, by a process of elimination, the speculation would eventually narrow to the point when there would be no one left to be suggested as an alternative apart from the real author himself. "You see Waring turns out not to be the author," Seeley wrote to Sidgwick on 15 May: "the suspicion of me does not seem at any rate to have got into print yet."[16]

By May, Seeley had indeed avoided being connected with the book in print, but his name was circulating, along with a set of other possible candidates, in less formal settings such as at bookshops and dinner parties. The rumour of Seeley as a viable candidate was, perhaps unsurprisingly, first taken seriously in Cambridge, where Seeley had been an undergraduate and later fellow of Christ's College. As this news began to reach Seeley, Macmillan sought to reassure him that his name "is only known or guessed in one set at Cambridge. I have seen several Cambridge men lately who don't guess in the least." Macmillan was adamant that Seeley should "not give in as yet ... Stanley is anxious that the secret should be kept."[17] Later in the month Macmillan tried to convince Seeley that the speculation in Cambridge was a minor setback, that his secret was still safe: "There is not the least danger of your secret. Guesses centre on the Johnson of Eton at this present. Edward Bowen was the previous favourite. Your name was uttered a little in Cambridge but incredulously and mildly. You are quite safe."[18]

Seeley was, however, concerned particularly about a few individuals who were likely to be convinced about his authorship, and in particular an old Cambridge friend of his, one Henry Sidgwick. His concern was confirmed by Macmillan in a letter that is worth quoting at length:

> I was down at Cambridge on Saturday and dined with Henry Sidgwick at Trinity and afterwards went to his rooms. He had read the book and was talking about it. He did not like he said the first part but the second was "the most invigorating reading he had had for long," it was "like drinking the priests strongest wine." He talked a great deal about it, and curiously enough he made the remark which you feared he would make. That he had heard *you preach a Sermon* one evening in his rooms, I think he said, which was very much like one of the chapters. I looked quite innocent, and said I had seen you lately and actually given you a cold, but had not heard you explain any opinion of the book. I don't think that he could possibly conclude anything from what I answered or said. But I confess that I felt more than half inclined then and there to take him into confidence, as the safest way towards shutting his mouth. But I am not sure. He did not seem to dwell on it afterwards, or return to it, else I think I would. But he may think all the more, and I would leave it to your judgment to do what you think good.[19]

Seeley hoped that he and Macmillan could continue to keep Sidgwick in the dark, and on at least one occasion Seeley declined to attend a gathering that would include Sidgwick, likely in order to avoid being put in a situation where he might have to confess his authorship or worse – explicitly lie about it.[20]

By early May, Seeley gave in to Macmillan's suggestion, thereby letting Sidgwick into what was at the time still a small circle of knowing friends. In one sense, this was clearly a positive development for Seeley. Sidgwick did indeed "shut his mouth," as Macmillan suggested he would, and he seemed to enjoy, as he said, having to "weave fraud and deception to the best of my ability."[21] As we saw in chapter 7, he became an important correspondent, with whom Seeley could discuss openly some of the problems of *Ecce Homo*, particularly in the context of adding the preface for the fifth edition. These conversations, therefore, did not need to be mediated through Macmillan but could proceed among knowing friends. Perhaps more importantly, Sidgwick was also able to act as another set of eyes and ears for Seeley about the secret authorship. He let Seeley know, in much the same way as did Macmillan, about the state of his secret. Sidgwick informed Seeley about the gossip concerning possible candidates for authorship and let him know when Seeley's name was uttered in connection with the book.

Indeed, just as Macmillan was reassuring Seeley that his secret was safe, Sidgwick was giving Seeley more detailed information about who was guessing his name and spreading the rumours of his authorship in Cambridge. "Lightfoot I fear knows," Sidgwick wrote to Seeley in reference to their mutual friend Joseph Lightfoot, who was at the time Hulsean professor of divinity at Cambridge and chaplain to the prince consort and the queen. "He consciously refused to tell a large party of whom I was one, *who* the author was, but he let out that he was a friend of mine, and said we should all of us soon know."[22] This irritated Seeley because he had actually not told Lightfoot. Moreover, how would Lightfoot know whether or not Seeley was going to unmask himself? "I do not understand what right Lightfoot can have, even if he has been told himself, to say that we shall all soon know. I propose to fight a good deal longer yet."[23] Clearly bothered by this news, Seeley continued the conversation a few days later, writing to Sidgwick that "Lightfoot may after all have got hold of some false report. Why should he have said 'a friend of your's?' I am a friend of his … and have seen a good deal of him."[24] But at least Lightfoot was not saying who he thought the author was. The same could not be said for Seeley and Macmillan's mutual friend, the Broad Church theologian and future author of *Flatland* (1884), Edwin A. Abbott.

Much like Sidgwick, it was determined that Abbott must be brought in on the secret because he had clearly realized who the author was and, unlike Lightfoot, was quite willing to share that information. In response to a letter from Seeley that must have requested that Abbott keep the secret to himself, Abbott wrote that "I fear you are too late. As you had not trusted me with your surprising secret, I did not feel bound to conceal suspicions even though they amounted nearly to certainty." Abbott said that he was convinced not because he was told by anyone in particular but because "it agreed with much of your conversation and many of your letters, [and] the style resembled your inaugural lecture" at University College London on the importance of classical studies. Abbott was forced to confess that he told J. Llewelyn Davies, W.L. Clay, Derek Hudson, and his own wife. He tried to assure Seeley that they were not convinced by his arguments in favour of Seeley's authorship, though Abbott's wife accidentally saw Seeley's letter that confirmed the secret. "[A]s I had told her before I felt certain you had written it, I thought it the lesser of two evils to tell her the secret which she had by … then half guessed. Now if you can't forgive that, you are a brute and had better read the sermon of forgiveness which I shall shortly send you." Abbott went on to say that his wife "is silent as a rock in secret matter," so Seeley should not be concerned about her. Abbott also wanted Seeley to know that he would now keep the secret, given that he was officially informed of it.[25]

It was likely because of the cases of both Abbott and Sidgwick that Macmillan felt more inclined to take into confidence those who were fairly convinced that Seeley was the unknown author. On at least one occasion he did this without conferring with Seeley first. At some point in June he decided to tell David Masson, the editor of *Macmillan's Magazine*. "I knew I was imperilling no secret by telling Masson," he wrote to Seeley, "anymore than of whispering it to my wife – his reticence is boundless." Seeley was clearly unimpressed by this, as Macmillan was quite apologetic: "I will do nothing of the kind again without your leave." This was, Macmillan assured Seeley, a "special case."[26] Such special cases were multiplying, however, and it was only a matter of time before Seeley and Macmillan attempted to bring someone in on the secret who would simply decline to keep silent, not out of malice or spite but by embracing the very set of moral standards that were being promoted in *Ecce Homo*.

Indeed, even though anonymous publishing was a standard practice in the Victorian book trade, *Ecce Homo* was published at a time when that practice was being openly debated, particularly in regard to periodicals. While older quarterly reviews had traditionally published anonymous articles in order to promote a consistent editorial line that transcended the individual authors, new magazines and journals were eschewing this practice, thereby embracing a liberal individualist authorial identity quite in contrast to the corporate identity implied by anonymous articles.[27] Anonymous book authorship was not as widely debated, and engendered a related, though slightly different, set of problems largely surrounding notions of respectability and morality. It was paramount for public intellectuals to embrace a mode of comportment that was respectable; otherwise their views could be rejected as the product of a disreputable character.[28] Anonymity made it more difficult to judge the character of a person who remained hidden. What kind of person, for instance, would not stand up for his or her own public views? An otherwise positive review of *Ecce Homo* made precisely this point, arguing that "We own … to a prejudice against a writer who proposes to make a revolution in theological thought, but who is lacking in the necessary courage to affix his name to a book. We doubt whether any great revolution in thought has been effected by a writer who will only fight when he is covered by a shield which renders him invisible to his foe."[29] It was also widely assumed that anonymous authors would have to engage in suspect moral activities in order for their identities to remain secret. Dishonesty was likely the worst insult that could be directed at public intellectuals, particularly those writing about religious topics, if the debate between Kingsley and Newman is anything to go by,[30] as well as the controversy engendered by the "dishonest" authors of *Essays and Reviews* (see chapter 1). Anonymous authors were therefore caught in a troublesome dilemma when choosing to remain invisible, because doing so certainly entailed deception if not outright lying.

The *Saturday Review* offered commentary on this particular dilemma of anonymous publishing by directly referencing the speculation and probable deception associated with *Ecce Homo*. Suggesting that there were "white lies" that society deemed perfectly acceptable, such as the footman who says that his master is not at home, the *Saturday* argued that the real concern was "how to deal with a large number of polite fictions which are not so much white lies as whiteybrown ones." The example the *Saturday* provided of just such a "whiteybrown lie is the answer which an anonymous author occasionally gives to curious or intrusive acquaintances who interrogate him point-blank about his secret ... The British public, for example, at the present moment is very anxious to discover the author of *Ecce Homo*, and any one who might be [the author] ... has probably been asked at least ten times over, by different people, whether he is the unknown theologian." In such a situation the author could take the high ground and state that he will not answer the question, thereby effectively confirming the questioner's suspicions, or else he could follow Walter Scott's model and lie under the assumption "that deceit may be justifiable if it is a means to a good end." The *Saturday* argued, however, that this rationale "is inconsistent with the view that truth is to be considered as an absolute and inflexible rule of life; and the worst of it is that, if we admit of one exception to the rule of inflexibility, it is difficult to see where we are to stop." Accepting such "whiteybrown lies," therefore, was a slippery slope: "If a white lie is excusable which is told to keep the authorship of *Ecce Homo* dark, what are we to say of lies told, not merely for a personal convenience, but for some object of still more paramount importance, such as the good of the nation, or for the good of a church, or to save men's souls?" The *Saturday* continued: "Before we know where we are we find ourselves on the confines, if not of Jesuitical casuistry, at all events of the position which [William] Paley certainly upheld, but which modern moralists usually condemn."[31] While the article in the *Saturday* not only put the general dilemma of anonymous authorship in stark terms, it also specifically suggested that the only possible way for the secret of *Ecce Homo*'s authorship to have been maintained, thus far, was through outright deception and lies by its masked author. Seeley immediately grasped just what was being implied about his moral standing. He wrote to Sidgwick about "the Saturday's attack upon me as one who *must* be telling lies."[32]

What the *Saturday* failed to mention, however, was that it was considered impertinent to query an anonymous author openly about his suspected identity. This meant that it was in theory unlikely that an author would be in a position to have to deny something explicitly and could therefore follow a series of evasions to avoid answering less direct questions. It seems that this is precisely what Macmillan and Seeley did. A useful example appears in a letter Seeley wrote to Henry Allon, the editor of the *British Quarterly Review*. At the time it was being suggested that the author of *Ecce Homo* was a Unitarian, a view that Allon must have been perpetuating. Seeley, of course, acted as if he did not know who the author was, but at the same time he did not want his book too closely associated with a particular Christian denomination. This was a suggestion, therefore, that he was particularly concerned to deflect. "Whether the writer is a Unitarian or not, nobody can tell," Seeley wrote; "there are passages in the book which look as if he was not, and there is certainly nothing which proves that he is." Seeley then mentioned that he had heard of one instance where a man was converted *from* Unitarianism thanks to the book. With that said, however, Seeley argued that he believed that the debates about the author's identity were largely irrelevant. "I think it should be read simply for what it professes to be, an Essay on Christian Morality. As such I like it; it has a great deal of freshness and spirit, and a novel way of putting things." But then he concluded his letter by offering his own speculations as to the author's identity: "The writer does not strike me as exactly learned, but well-informed and cultivated. It has been attributed to everybody from the Emperor Napoleon down to me. I fancy this last notion was suggested by Macmillan's saying that the writer had been at two Universities, which, though it is not true of me, might be supposed so, because Univ[ersity] Coll[ege] is constantly mistaken for a University."[33] Note that Seeley does not deny that he is the author, but instead problematizes the evidence that had been mobilized in favour of his authorship. This no doubt provides some insight into how he might evade any direct questioning.

Indeed, it seems that neither Macmillan nor Seeley ever outright denied the possibility that Seeley authored the book, though in such situations Macmillan would suggest other possible authors as well. As the speculation began to centre on Seeley towards the end of June, just after a short note in the *Patriot* had suggested that he could be the author (see chapter 8),[34] Macmillan wrote to Seeley that "I quite agree with you that just at present we must be careful. I think I always steadily acknowledged that you are one of the candidates and naming a lot with you we may still keep up the uncertainty, which will do quite well for our purpose."[35] It also seems that Macmillan was naming these other possible candidates under the premise that he too was not in on the secret and therefore was himself only guessing. This, no doubt, would have simply been considered by the *Saturday* as another kind of "whiteybrown" lie, one no better than an outright denial.

There was at least a vague code that Macmillan and Seeley were following when it came to evading questions about an unknown author's identity, but the same was not the case for an author's knowing friends.

There was no code of conduct that protected associates from being asked directly if they knew the secret or not. And those questions no doubt led precisely to the double-bind that the *Saturday* noted with regard to authors being directly questioned. We have already seen that Sidgwick, at least initially, was not terribly bothered by the necessary deception and lying that went along with being a part of Seeley's secret. The Cambridge philosopher John Venn also does not appear to have minded being asked by Seeley "to throw any inquisitive people I might encounter off the scent." He claimed that he did his best to follow Seeley's directives. Whenever someone began to get "warm," Venn would say "that I had seen so much of Seeley, and talked so freely with him on cognate subjects, that had he been the author I must certainly have known."[36] Abbott also suggested that he would do what was necessary in order effectively to keep the secret. "And lie I will right and left if I can keep your secret," he assured Seeley. But he wanted Seeley to know that he could only rationalize such activity by "consider[ing] the lies yours and not mine for you have forced them on me by not trusting me from the first."[37] Indeed, "whiteybrown lies" were being relied upon to keep Seeley's secret, whether Seeley was himself directly lying or not.

This was precisely the problem J. Llewelyn Davies attempted to articulate in a long letter he wrote to Macmillan on 13 November 1866 (see figure 9.2) that sought to justify why he was speaking "without reserve" about Seeley's authorship. Davies was part of a circle of Cambridge friends that included Seeley, Abbott, and Macmillan. Davies was, moreover, one of Maurice's closest disciples. He was ordained a priest in 1851 and in 1856 he became vicar of Christ Church, Marylebone. This was the same year that his critique of Jowett's commentary on St Paul was published by Macmillan in Cambridge (see chapter 1). Given that Davies was well integrated into this close-knit group of liberal Cambridge Anglicans, it was not difficult for Abbott to convince him that Seeley was the great unknown author of *Ecce Homo*. But, unlike Abbott, Davies was not sworn to secrecy. He moreover refused to hold his tongue because he regarded the anonymous publishing of *Ecce Homo* and the way its anonymity had been protected as unsavoury. "In conversing with you about the authorship," Davies wrote to Macmillan, attempting to explain his actions, "I had always said it was inevitable that it should become known. When I was mentioning to you that in my belief it was known, I could not help thinking of the way in which the secret had so far been partially kept. I know it would have come out long ago, but for direct denial. I said this denial gave some pain to me

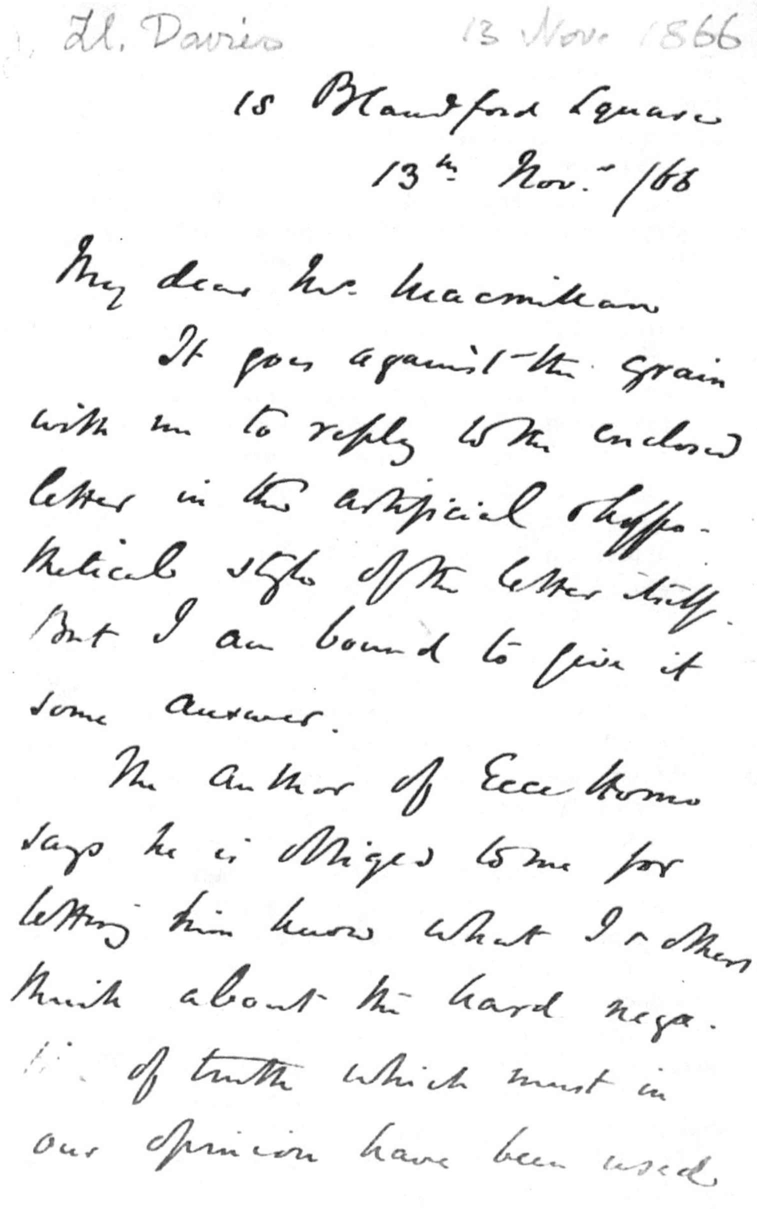

Ll. Davies 13 Nov. 1866

18 Blandford Square
13th Nov. /66

My dear Mr. Macmillan

It goes against the grain with me to reply to the enclosed letter in the artificial & hypothetical style of the letter itself. But I am bound to give it some answer.

The Author of Ecce Homo says he is obliged to me for letting him know what I & others think about the hard nega- [illegible] of truth which must in our opinion have been used

Figure 9.2 J. Llewelyn Davies to Macmillan, 13 November 1866, Seeley Papers MS 903/3A/2. Courtesy of the Senate House Library, University of London.

and others." Davies explained that Macmillan and Seeley clearly did not understand the difficult positions they were putting their friends in by engaging in evasions while their friends were expected to outright deny when questioned about the authorship. He believed that Seeley had already effectively admitted authorship by refusing to deny it himself on several occasions. Given that "the authorship ... was really known without any manner of doubt to an indefinite number of persons," there was no good reason for Davies himself to lie or engage in similar evasions when he became convinced of Seeley's authorship. "I have been constantly asked if I knew who wrote Ecce Homo, and I was aware of no sufficient reason to induce me either to tell a falsehood about it, or, by speaking evasively, to lead even stupid people to fancy I had written it myself." Davies believed himself to be taking the moral high ground and he expressed "regret that this noble and beautiful book should have this kind of association connected with it. *Ecce Homo* will become *the* instance in the casuistry of anonymous publishing," he wrote. "Indeed, it has already been spoken of in an article in the Saturday Review as a presumable case of denied authorship."[38]

What Davies failed to admit in this letter was the fact that one of the people he had recently spoken to about Seeley's authorship was R.H. Hutton, the editor of the *Spectator*. Whatever information Davies presented to Hutton was clearly convincing: the 10 November issue of the *Spectator* declared that "*Ecce Homo!* appears at last definitely traced to Professor Seeley, of University College, London." The notification went on: "The author complained in his recent preface of its being supposed that he could wish to mystify the public as to the drift of his treatise. We suppose he felt no scruple as to his authorship, as he seems to have succeeded admirably in mystifying even intimate friends."[39] Of course, it was precisely because of one of Seeley's "intimate friends" that his name was now definitively connected with the book in print. Making matters worse was the fact that a week later other weeklies reported the *Spectator*'s declaration as fact. The *London Review*, the *Athenaeum*, the *Reader*, and *John Bull* reported Seeley as the author.[40] The *Reader* was even helpful enough to include a short biography of Seeley, writing that he "is a son of Mr. Seeley, the well-known Evangelical publisher, and was educated at the City of London School, under Dr. Mortimer ... He is Professor of Latin in University College, London."[41] News of Seeley's authorship also spread to newspapers such as the *Pall Mall Gazette*, the *Bury and Norwich Post*, the *Leeds Mercury*, the *Bradford Observer*, the *Ipswich Journal*, the *Leicester Chronicle*, and *Berrow's Worcester Journal*.[42]

By 18 November, the *Era* announced that it had been a full week since Seeley was declared the author and there as yet had been no statement given to suggest otherwise.[43] Nor would there be.

In response to these public outings, Macmillan seemed relieved. "I suppose I need make no mystery now, need I?" he wrote to Seeley on 12 December.[44] There was, according to Macmillan, little point continuing to deny the speculation. Surely now was the time to stand up and confirm what had already been widely reported.

Chapter Ten

Behold the Man

Let no one ever trust *me* with a secret; I shall vow secrecy and then instantly proclaim it to all the four winds by way of revenge for what I have suffered.

– J.R. Seeley to Joseph Mayor, 26 November [1866]

Understandably, Seeley and Macmillan were angry to learn that their secret had been exposed by an "intimate friend." Macmillan wrote to Davies to voice their displeasure while making it clear that they found his justification for telling Hutton the secret completely untenable. "The moral indignation which drove you to come and chide me and the author and proclaim us in the Spectator," Macmillan wrote to Davies, "was an indignation founded on very imperfect knowledge – pardon me in saying it was an unjust indignation." Macmillan admitted that there were a few occasions where Seeley's previous evasions could have been construed as admissions, though he said that Davies was exaggerating about the effect those evasions would have had on the overall state of the secret, which was far from being discovered. But this was really beside the point, Macmillan argued. "Altogether apart from other considerations, in a case of this kind where a man deliberately decided that he wishes to be anonymous I do not think his wish ought to be tampered with. If it were a mere whim, it ought to be gratified. If it were not known to be a whim, other and better mention might charitably be conceded to him. I think in any case he ought to be protected not proclaimed."

Macmillan was also irritated with Hutton, whose explanation for "proclaiming" Seeley was equally inadequate. "I think you and he [Hutton] ought to do your best to heal the breach you have made. What your conscience will impel or permit you to do, I leave to yourselves."[1] A month after this exchange, Macmillan and Seeley were still angered by the actions of Davies and Hutton. "I will write Davies as you wish," Macmillan wrote Seeley on 12 December 1866. "I think he deserves it." He continued, "The impertinence of the Spectator was great and *little*. But Hutton is a pachyderm. Alas for us thinner skinned critters."[2]

Despite Seeley and Macmillan's anger about Davies and Hutton divulging Seeley's secret, there was some validity in Davies's rationale for refusing to lie on someone else's behalf. From Davies's perspective, he was merely speeding up the process in order to save everyone the indignity of continuing to feign ignorance. Not only was it immoral to engage in evasive behaviour to protect Seeley's secret, according to Davies, it was unnecessary, given that the secret was known by an indefinite number of persons. As Davies suggested, the way the secret was half kept meant that it was inevitable that it would eventually come out.[3] While this may be true in the case of *Ecce Homo*, it raises a more general question about the viability of anonymous publishing: Was it possible for the author of a literary sensation to remain anonymous in this new era of modern communication? Of course there were many other anonymous books that were published in this period that could be examined to answer this question. In terms of popularity and content, however, the publishing of Robert Chambers's anonymous *Vestiges of the Natural History of Creation* offers a very useful comparison to that of *Ecce Homo*. And it suggests that it was indeed possible to publish a sensational book about a highly contentious subject over the space of several decades without having the secret of authorship become widely known. But doing so required a much more thorough and well-planned strategy of anonymous publishing than that carried out by Macmillan and Seeley.

Robert Chambers and his publishing partner, who happened to be his brother, were very careful about whom they let in on the secret of Robert's authorship. Robert could certainly trust his brother, William, with keeping the secret because the respectability of the Chamberses' Edinburgh publishing house would have likely been damaged by the suggestion of Robert's authorship. Such a revelation would have hurt William as much as Robert. Other than William, only a handful of trusted close friends and family members knew the truth. Even so, Robert made a mistake in allowing at least one person to discover the secret who could not be trusted: the editorial assistant of *Chambers's Journal*, David Page. When Page quit the publishing firm after being denied partnership, he went public with his secret in the *Athenaeum* on 2 December 1854.[4] Luckily for Chambers, Page was explicitly named as the source for the rumour along with what appeared to be quite convoluted evidence for

Chambers's authorship. Page was widely viewed as a disgruntled former employee looking to get even with his well-respected bosses, so his claims were largely ignored.[5] However, this incident was a harsh reminder to the Chambers brothers that very few individuals could be trusted with such knowledge, and they kept the list of people in the know to a few who would never confirm the secret.

This is not to suggest that Chambers was not suspected of being the author of *Vestiges*, because he certainly was. But so were many others, including one Charles Darwin. As long as Chambers continued to deny authorship and as long as anyone credible who had proof of the secret continued to feign ignorance, the mystery of authorship would remain, despite the growing speculation that Robert Chambers was indeed the author. Most important, however, was that Chambers did not have to worry about whether or not his publisher was able to dodge questions about his authorship. Unlike Macmillan, the publisher of *Vestiges* well and truly did not know who the author was.

The fact that Chambers was able to publish *Vestiges* without his publisher's knowledge of authorship constitutes the greatest distinction between the publishing of *Vestiges* and that of *Ecce Homo*. *Vestiges* was published by John Churchill, who typically published medical texts. Chambers dealt with Churchill through an intermediary: his trusted friend Alexander Ireland. For his part, Ireland avoided any face-to-face contact with Churchill, and all documents written in Chambers's hand were "recopied by Ireland before being forwarded on to Churchill." Such communication, according to James Secord, "was cumbersome but effective." Moreover, communication between Chambers and his publisher was filtered through intermediaries in both directions, thereby keeping both publisher and author well removed from one another.[6] Not only did this mean that there was very little evidence connecting Chambers to Churchill, it also meant that Churchill could genuinely claim ignorance when being questioned about the author's identity. Any suggestion on his part about who the author might be would have been pure speculation.

This distance also precluded any divergent interests between author and publisher coming into play, at least when it came to the issue of maintaining anonymity. Churchill had no choice but to follow the directives of Chambers's intermediary – unless of course he wished to discontinue publishing *Vestiges*. It goes without saying that Macmillan and Seeley had a much less formal arrangement. Even though Macmillan at times suggested that he did not know the author, he was not consistent in doing so, as Rose Kingsley's note to him (see chapter 9) makes clear.[7]

Moreover, Macmillan denied that certain people such as Goldwin Smith were the author, undermining his claims that he did not know who the author might be. It is true that, like Seeley's, Robert Chambers's identity was guessed at by quite a few people, yet he engaged in a much more thorough and complex process of anonymous authorship. So while Chambers's readers could never be entirely sure about the author's identity, readers of *Ecce Homo* were convinced that they knew, in large part because there was a direct connection between author and publisher that could not be denied by readers seeking out *Ecce Homo*'s secrets. There was also, however, a central miscommunication that helps to explain Macmillan's muddled strategies for maintaining *Ecce Homo*'s anonymity.

Macmillan seemed to be operating under the assumption that the secret of *Ecce Homo*'s authorship would eventually get out and that Seeley would want to sign his name to a new edition once this occurred, following a familiar nineteenth-century publishing practice of exciting interest in an unknown author's first book.[8] *Ecce Homo* was, after all, Seeley's first book and within a year of being published was undoubtedly a success. It does not appear to have occurred to Macmillan that Seeley's authorship would remain a secret in perpetuity. That Macmillan was proceeding under this assumption is apparent from the very beginning. In his letter to Gladstone, for instance, Macmillan states that the author has chosen to maintain anonymity "*in the meantime*," a phrase that is significantly underlined in the original letter, though not in the copy that was sent along to Seeley.[9] Moreover, in his correspondence with Seeley throughout the year, he often suggested that Seeley should "not give in," as if it was a foregone conclusion that he eventually would. "I certainly would not give in *as yet*," Macmillan told Seeley on 22 May, when the sales were still going strong.[10] Moreover, others were clearly working under the same assumption. We have seen that Davies assumed that the secret would eventually get out. Stanley also assumed as much but wanted the name of the author not to "be disclosed till after the appearance of the 2d volume."[11] Given that Macmillan was operating under the assumption that the secret would eventually be known, his strategy for increasing speculation about the author makes sense. It was a decidedly short-term strategy that would increase interest in the book, though in the long term it would inevitably lead readers to the most likely author. Which is exactly what happened.

When the secret was widely publicized, thanks to the *Spectator*, in early November, most assumed that Seeley would, in some form or other, publicly acknowledge authorship of the book. "Are you going to give in or not?" Sidgwick wrote to Seeley on 15 November. "I only want to know when you think the game up. The complete apathy with which the announcement in the Spectator has been received here, at least as far as my observation has gone, I can only account for by supposing that the secret has permeated society here to a considerable extent."[12] That the secret had "permeated society … to a considerable extent" was clearly Macmillan's view as well. Less than a month later, he was ready to unmask his author. Even though he was angry with Davies and Hutton, he was also relieved that he no longer had to engage in all the acts of duplicity that were required to keep up the ruse. He asked Seeley if he could officially disclose the secret, which he was still somewhat hesitant to do. "I have been still saying that I was not authorized by the author to say who it was. If you authorize me to use my liberty I will do so."[13] What Macmillan failed to realize, however, was that this was a secret that Seeley never intended to admit publicly, even when that secret became one in name alone. Seeley was simply never prepared to grant Macmillan the authority to do so. Indeed, Seeley's name would not appear on an edition of *Ecce Homo* until the posthumous edition was published, in 1895.

Depending on the person, however, Seeley was quietly admitting that he was the author. For many, the press announcements simply confirmed what they already believed, so he had little choice but to confirm the rumours in such instances. W.H. Parry wrote to Seeley to say that "your name is in everybody's mouth or, at any rate, the name of the book of which you are supposed to be the author." Parry claimed that he "felt a little proud of numbering among my friends one whom I believed rightly regarded as the writer of 'Ecce Homo.'"[14] J. Skelton wrote to Seeley in order to "renew the friendship which the altered position you hold towards the world might be thought to have impaired. I allude of course to your being, as it is reported, the author of 'Ecce Homo.'" Skelton wanted Seeley to know, however, that he was one of the book's early supporters. "I assume that the report of the authorship is true; if you do not care to acknowledge or disclaim it, don't take any notice of what I have written!"[15] Frederick Fitch wrote to Seeley that "I was much surprised yesterday by seeing a paragraph in the Guardian quoting from the Spectator, ascribing the authorship of Ecce Homo to you – is this correct?"[16] Unfortunately, Seeley's responses to these correspondents are no longer extant, but a similar exchange with his friend Joseph Mayor has survived and perhaps gives us a sense of how Seeley responded to these queries about his identity in the wake of his secret being exposed.

Mayor was one of Seeley's Cambridge friends. A philosopher and classics scholar, Mayor entered St John's College in 1847, received an MA in 1854, and was ordained a priest in 1860. Along with Seeley, his network included Abbott and Sidgwick. He was also very close with Lightfoot.[17] As we will see below, Seeley came to believe that Mayor was the mutual friend Lightfoot claimed as an authority when he said that he knew who the author of *Ecce Homo* was in a conversation with Sidgwick, a conversation Sidgwick relayed to Seeley.[18] That episode had occurred back in May.

It was now November, and Mayor had his assumption much substantiated. After seeing one of the announcements in the *Pall Mall Gazette*, Mayor wrote to Seeley that it had also been "confirmed by Abbott ... that you were the great unknown. How you must have chuckled at the blindness of your friends."[19] Seeley responded in what can only be described as a resentful manner. "I am tired of evasions," he wrote to Mayor on 26 November (see figure 10.1), "and I have not for a long time believed that I could keep the secret from you. I am obliged to give up, but it is a sullen submission." There was also some anger that was barely concealed in the letter, however, finally giving us a glimpse of how he felt about the seemingly careless way his secret was kept by both his publisher and friends. "I am angry with people for their reckless curiosity and with some of my friends Abbott, Macmillan, [W.B.] Gunson, for their utter inability to hold their tongues. Let no one ever trust *me* with a secret; I shall vow secrecy and then instantly proclaim it to all the four winds by way of revenge for what I have suffered." He was thankful that Mayor had been much more discreet than his other friends, though even in Mayor's case Seeley was sure that he was not entirely blameless. "I believe you have known it for a long time. The last time I dined with you – by the way, it was black Friday – I thought you knew it and came to my rescue when the subject was broached. You have been much more considerate than most people and I thank you for it, but I believe you told Lightfoot; at least I cannot account otherwise for his having guessed so early."[20]

What Seeley's friends may not have entirely recognized was the fact that perhaps Seeley wanted to remain anonymous for a practical reason – he did not want his evangelical family to be associated with a book that he knew would likely scandalize them. Unfortunately for Seeley, by being so publicly proclaimed in the weekly press, it was no longer his university friends that were speculating about his authorship but his family members as well. "I receive now," Seeley told Mayor,

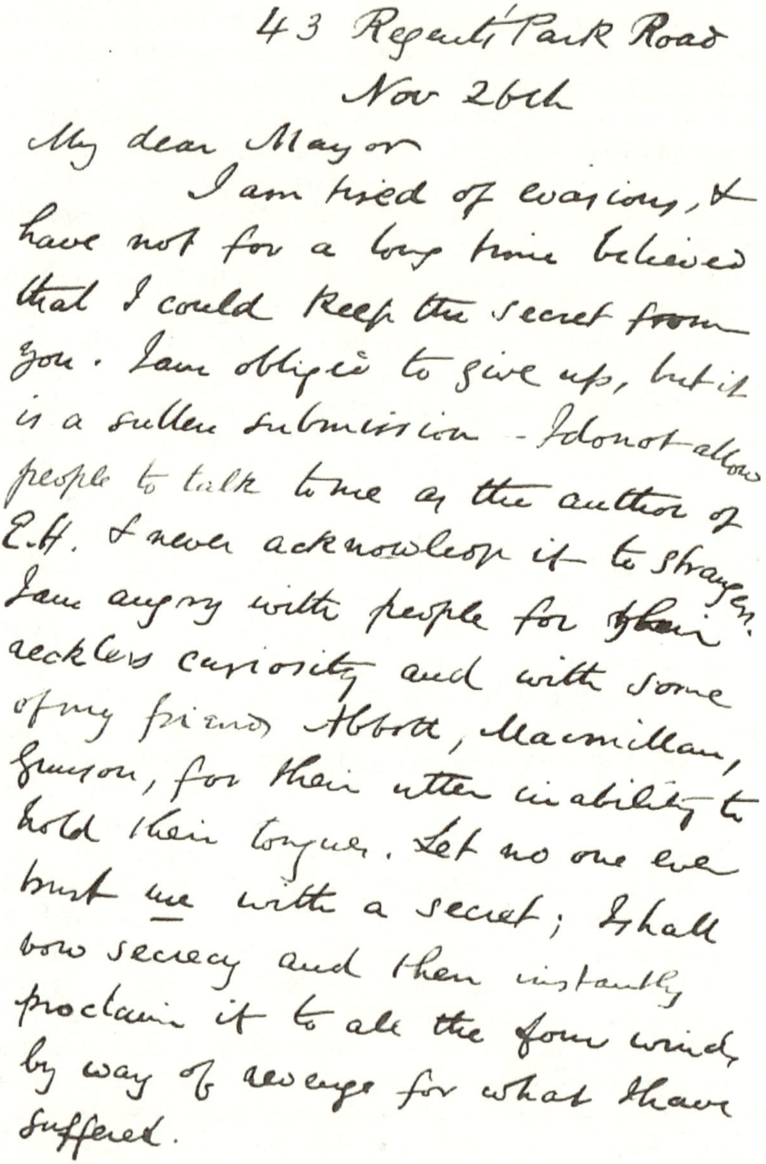

43 Regent's Park Road
Nov 26th

My dear Mayor

I am tired of evasions, & have not for a long time believed that I could keep the secret from you. I am obliged to give up, but it is a sullen submission - I do not allow people to talk to me as the author of E.H. & never acknowledge it to strangers. I am angry with people for their reckless curiosity and with some of my friends Abbott, Macmillan, Gunson, for their utter inability to hold their tongues. Let no one ever trust <u>me</u> with a secret; I shall vow secrecy and then instantly proclaim it to all the four winds by way of revenge for what I have suffered.

Figure 10.1 Seeley to J.B. Mayor, 25 November 1866, Seeley Papers, MS 903/1A/1. Courtesy of the Senate House Library, University of London.

"from relatives and others – letters claiming me with Judas Iscariot."[21] A case in point was a letter Seeley received that was dated 26 November, the same day that Seeley sent his note to Mayor. "There is a sad rumour about you," Seeley's cousin Maria wrote, "which being uncontradicted, seems to be too true." Because the two of them shared the same "name and baptism," she urged him "to 'reconsider' the awful step you have taken." Seeley had dishonoured "our Lord and Savior" and had no doubt caused "trouble [to] enter many hearts and homes." "I fear you have gone to Milton and others, rather than to Matthew, John, Isaiah, or Paul, to learn how Paradise is regained." She pleaded with him to "use not for Satan those flowers which might make you a standard-bearer in Christ's army!"[22]

Someone in the family had also been educating Seeley's younger sister, Anne, about the evils of *Ecce Homo*, though it seems Anne was unaware that Seeley was the author. "You are really alarming yourself needlessly about Ecce Homo," Seeley wrote to Anne on 25 November. "It seems you have not seen it, but have been told by people that it is very heterodox … The book may very likely be unsettling to people who are quite happy in orthodoxy. I do not by any means advise you to read it. Evidently it was not intended for such persons, but for sceptics and disbelievers. The writer gives fair notice of this and it is their own fault if the wrong set of people read the book after this … Yet you think the book one not to be published, because it may unsettle some people who had no business to read it at all. Is not this rather selfish of you, my dear Anne? One more unsettling book can make little difference now, one book that establishes something and gives people something to stand upon may make great differences."[23] This kind of conversation was precisely what Seeley wanted to avoid by publishing anonymously. Therefore, even while his secret was fairly well known by the end of 1866, it was important for him to maintain the now very thin veil that remained between him and the book.

It has often been speculated that Seeley elected to publish *Ecce Homo* anonymously to avoid offending his father, who, as we have seen, was active in the evangelical publishing world and vocal about his opposition to many of the streams of thought that influenced his son (see chapter 2). There is no doubt that R.B. Seeley would have known about *Ecce Homo*. Indeed, he very well could have been at the May meeting of the Church Pastoral Aid Society and witnessed first-hand his friend Lord Shaftesbury denounce *Ecce Homo* as a regurgitated product of Hell. But as we have also seen, the elder Seeley was familiar with his son's views, as they had long debated the merits of Broad Church theology (see chapter 3). There is, unfortunately, little evidence to confirm whether or not he was angered by his son's decision to publish them in such a manner. It does seem significant that he doesn't appear to have written about *Ecce Homo*, even though he was at roughly the same time writing evangelical polemics against *Essays and Reviews* and Bishop Colenso's works. He was also silent when it came to addressing his son in his will. J.R. Seeley was the only one of R.B. Seeley's children not named as a beneficiary upon his death in 1886.[24]

This is all to say that there were clearly familial consequences for Seeley in having his name so openly associated with *Ecce Homo*. So while he may have been quietly admitting in private that he was the author, he did not want to make a public admission of this fact. He made this clear to his friends, who were no doubt disappointed to learn that they still could not speak freely about the book. He explained to Mayor, for instance, after the secret was well known, that he was refusing "to allow people to talk to me as the author of E[cce] H[omo]. I never acknowledge it to strangers."[25] And this was a practice he apparently continued. A full year and a half later, after the *Spectator* had named him as the author, he still instructed his friends to avoid discussing the book or its authorship in his presence. In response to an invitation to breakfast with Gladstone, for instance, Macmillan expressed Seeley's wish to attend but wanted Gladstone to know that Seeley was still rather touchy about one subject in particular. "Might I venture to say that Mr Seeley has not been addressed in general society as the author of E[cce] H[omo]," Macmillan wrote to Gladstone, "and is anxious to avoid even the discussion of the book in his presence. I am sure you will understand his feeling on the subject."[26] This letter was dated 25 June 1867.

By then, however, interest in *Ecce Homo* had fallen considerably, going some way to prove Macmillan right that once the speculation about authorship stopped, so too would the sales.[27] However, a cheap edition of 3,000 copies was published in January 1867, followed by another 3,000 copies printed in March.[28] This brought the total sales figures up to about 16,500, which was very good indeed for a book that was only sixteen months old. But the sales slowed considerably after the cheap editions were published. A spike of interest was renewed in early 1868 when Gladstone published a series of essays that were later in the year published as a book entitled *On Ecce Homo*.[29] Without stating Seeley's name, Gladstone followed closely the opinion of Stanley, arguing that it was best to divorce *Ecce Homo* from the identity of any supposed author. But after initially summarizing some of the key arguments of the book, Gladstone set about outlining his own theological views concerning Christ's life, meaning that *Ecce Homo* was really the hook rather than the subject of the essays, which was a point of irritation for several reviewers.[30] Seeley was still appreciative of Gladstone's extended meditation on his book, and Macmillan relayed to Gladstone that Seeley believed he was "materially strengthening the position he [Seeley] sought to establish in this great field of action and thought."[31]

Gladstone's piece also reminded the general reader that there was still a companion volume to be published by the author of *Ecce Homo*. But that was a study that Seeley had decided to put off indefinitely. Macmillan was anxious that it be written and published to take advantage of the interest in the subject that was generated by *Ecce Homo*, but Seeley was no longer as motivated to articulate the theological views that are only alluded to there. During the height of the controversy surrounding the publication of *Ecce Homo*, he spent much of his energy working on an annotated edition of Livy, which included a lengthy "historical examination," for Oxford University Press.[32] As Macmillan was, at the time, the publisher for Oxford University Press,[33] he also had a vested interest in Seeley completing the project. He therefore thought that Seeley was "right to work away at Livy, and to keep your mind easy."[34] But as Seeley's Livy project extended into the next year, Macmillan became more and more anxious for Seeley to complete it.[35] "I shall be very glad to see the Livy Essay when it accomplishes itself," Macmillan wrote to Seeley on 27 March 1867, "and so will my Oxford masters."[36] No doubt, Macmillan hoped that soon after Seeley finished his study of Livy he would return to the promised sequel to *Ecce Homo*. He eventually would, but not for several years.

This is not to suggest that Seeley stopped thinking about theological matters or even writing and talking about them. In 1867, for instance, he read a paper before the Cambridge University College Students' Christian Association on the general subject of Christian faith. The paper gives clear evidence that, even while Seeley may not have wanted to have his name so closely associated with *Ecce Homo*, he did not shy away from sharing some of the views that were contained in the book. It also gives us a sense of how his religious views may have been changing in light of the reception of *Ecce Homo*.

In the paper, Seeley argued that "[t]here are two sorts of Christian faith." The first sort of faith was the kind that "abides by the principle of authority, and is before all things obedient and humble."[37]

Those that adhere to this kind of faith, argued Seeley, tend to be exceedingly loyal to the Church and revere the historical institutions. They have no desire to censor or condemn those of other faiths, but they also do "not have the courage to be tolerant and comprehensive." This was a kind of "unquestioning conservatism" that was certainly necessary for many who are "not able to assimilate new truths." There was nothing necessarily wrong with this kind of faith, argued Seeley; "we are not all of us made for inquiry."[38] It was a faith of feeling rather than a faith of thought, and he stressed that the Christian Association would be fortunate to have men of this kind within its ranks.

But there was another kind of faith that was also necessary, argued Seeley, one that directs men to "think and criticise, welcome rather than shun what is new, and find a sacredness in the present as well as in the past, and feel a Spirit speaking in them as well as in holy men of old." It was to this "class of men" that Seeley sought to direct the majority of his comments. This was because, as he said, it was currently "a critical time in the life of the Church." Orthodoxy was being assailed. And foundational doctrines were being challenged. "Never was there so universal a rebellion against the influence of the clergy."[39] This was happening, moreover, at a time when the principles of the New Testament were being taken up "far beyond the limits of any ecclesiastical organization."[40] But those outside the Church, who embrace its founding moral philosophy, claim that the Church has nothing to offer for the problems of the modern world, and that it should therefore be abandoned.

Seeley admitted to sharing some sympathy with this view because he argued that the Church needs a "philosophy of society," which is not something that can be found in the Bible, given its historical character. "In the Bible you will find great and universal principles, but every age brings with it a new society, and demands a new teaching, a new philosophy."[41] Here Seeley was much more explicit about one of the subtle themes of *Ecce Homo*, that Christianity must adapt to changing historical contexts for it to survive and be useful. "There are two books in which the Christian must perpetually read," argued Seeley, "the Bible and the Time."[42] In order for Christianity to be a powerful force, it must therefore adapt itself to the conditions of the present. And for Seeley, this meant specifically adapting Christian principles, such as charity, to meet the necessities of the modern social world. It is for this reason that Christianity, more than anything, needs a philosophy of society, so that Christians can be expected to understand adequately how modern society functions and can therefore contribute to making it function better.[43]

In response to some written queries about the talk, Seeley expanded on a few points worth considering here. He stressed that he was not referring to a particular denomination of Christianity but in general argued that they all failed to provide a true philosophy of society. When one of the queries wondered if Quakers might be an exception to Seeley's critique, Seeley argued that, just like other denominations, Quakers "do not press evils to their causes, nor assail them at their root." And seeking out the true nature of social problems was precisely what the Christian Church ought to be doing. "I profess I hardly understand what the Christian Church is good for if it does not point out the evils earlier and more energetically than men of the world, and form a juster and calmer conception of the proper remedies." Seeley argued that the Church was uniquely positioned to take up the problems of the modern world and present idealized solutions to them. "I regard the Church as the ideal society existing within the real one, the future within the present, a sort of provisional vehicle for ideas before they have power enough to embody themselves in political and social institutions."[44] This was, indeed, a powerful and modern vision for the Christian Church, though Seeley stressed that "no Protestant school has ever risen to the dignity of this conception."[45]

At roughly the same time, Seeley contributed an essay to a collection called *Essays on Church Policy*, which was edited by his friend W.L. Clay and published by Macmillan. Davies and Abbott also contributed essays to the volume, which was as a whole ostensibly about the national character of the Church of England. Even though the very short preface followed the practice of *Essays and Reviews* by suggesting that each writer was responsible only for his particular essay, the volume was viewed as presenting a particular "Broad Church" line about the future of the Church of England that extended throughout the eight essays.[46]

Seeley's essay "The Church as a Teacher of Morality" was particularly explicit that it was the Broad Church clergy who offered the best hope for the future relevance of the Church of England, particularly in regard to the issue of morality. Unfortunately, Seeley argued, the vast majority of "the pulpits in England are in the hands either of men who are slaves to tradition [the "High Church"] or of men who divide their congregation [the "Low Church"]," which means "that the people of England are not taught morality at all."[47] But unlike the evangelical party, the Broad Church aspires to represent not a particular chosen group of Christian believers but the community itself. It is unlike the High Church, moreover, because it is not conservative. "It repudiates the principle of authority in the investigation of truth; and if it abides by some ancient beliefs, and would retain them as the basis of modern order, does so on the ground that they are true, and that they are the best and strongest foundation upon which modern order can be based." It was for these reasons that the Broad Church was well positioned to teach a form of "Christian morality suited to the age."[48]

In order to do so, however, Seeley argued that the Broad Church clergy needed "a large and deep intellectual cultivation ... Their learning must be the instrument and material of original thought, not the substitute for it; it must be a knowledge of modern affairs rather than ancient, of the present rather than the past."[49] This would, of course, necessitate education itself being reformed so that it makes central what Seeley called "the progressive spirit," meaning that such an education would provide "insight into the present age and the actual existing society."[50] The positivism of *Ecce Homo* becomes much more explicit here, as Seeley calls on the Broad Church to adopt a "progressive ethic," which accepts "movement as the law both of Churches and States."[51]

To give an example of how this new progressive education could help with the teaching of morality, Seeley argued that the clergy would then be able to provide modern examples of morality rather than relying on examples from the Bible. The current reliance on the Bible as providing a guiding morality, Seeley argued, is a real problem, because "[t]he men of the Bible lived, in the first place, in circumstances unfamiliar to us." Given the immense effort involved in reconstructing those circumstances and then properly drawing from them the particular moral exemplar that is then reinterpreted in a modern context, it would be far easier and more relevant to focus on modern history.[52] It was, therefore, only the Broad Church, with the right kind of education, that would be able to guide the spiritual future of the English nation in what was a rapidly changing society.

If, as the author of *Ecce Homo*, Seeley convinced many readers that he was largely above getting involved in the traditional disputes between familiar Church parties, this essay in particular suggested that he was throwing his lot in with the Broad Church. And as the "reputed author of *Ecce Homo*," as the *Saturday Review* put it, Seeley gained a fair amount of notice with his essay, since he was, along with Davies, a name that was generally known to the public.[53] The *Christian Remembrancer*, for instance, found that the volume was a "sequel" to *Essays and Reviews*, and included views that were "in advance" of its predecessor in the sense that the authors seem "to endorse the whole series of denials of doctrine which have recently become so famous." While being much harsher in tone about the other essays, the reviewer argued about Seeley's that it was written by a man who does not really understand the way in which traditional theological doctrines are brought to bear on modern problems of morality. "Has he really never heard humility preached in connexion with the incarnation; love to God enforced by what God has done for man; purity or unselfishness from the example of our blessed Saviour. Has he never heard of the fear of God as connected with eternal punishment as a deterrent from an immoral act?"[54] About Seeley's suggestion that modern history should be used instead of the Bible, the reviewer simply said that this sounded too much like Carlyle's famous dictum, "that every nation's true Bible is its history."[55]

The *Saturday Review* was also critical of the volume in general, but largely because it failed, in the reviewer's mind, to offer a conceivable scheme of reform. The reviewer found Seeley's essay "an exception in character to the rest," however, because it offered "a series of practical suggestions, many of them sensible and pointed enough, as to the best method of educating preachers for their office ... Few clergymen, of whatever school, could fail to derive many valuable hints from its perusal."[56] The *Spectator* also found much to admire in Seeley's essay, particularly the way in which it sought to inquire into "the best mode of introducing true popular life into the action of the Church." The reviewer admitted that he did not want to "speak too highly of Seeley's admirable essay" because of what it suggested about the Church's "neglect in not enforcing the highest *political* morality ... and of her deficiencies in inculcating the positive duties of modern life." With that said, the reviewer found Seeley's analysis completely justified.[57]

With these early post–*Ecce Homo* publications, Seeley was explicitly articulating a theme that certainly runs throughout *Ecce Homo* and yet is easily missed – about the need for Christianity to adapt to modern society. And, indeed, this would be a theme much more thoroughly worked out in *Ecce Homo*'s sequel. But, unfortunately, formalizing the views that were being articulated in these post–*Ecce Homo* publications would have to be postponed for the time being. This is because in 1869 Gladstone came calling with a request that would seemingly take Seeley in a dramatically new direction, well away from his work on establishing a Christian philosophy of society.

Chapter Eleven

Behold the Historian

When I first asked myself how this subject might best be treated, I remarked that it was one which especially needed to be gathered up in a person.

– J.R. Seeley, *The Life and Times of Stein* (1878)

It is difficult to understand how the author of Ecce Homo came to write in the manner of the Life of Stein.

– J. Llewelyn Davies to the editor of the *Spectator* (unpublished), 22 January 1895

In hindsight, the fact that Seeley was for most of his life considered a historian seems hardly a point on which to ponder. It goes without saying that *Ecce Homo* was a historical study that focused particularly on the specific historical circumstances of Christ's life and early Christianity. Moreover, even his shorter pieces on Christian faith and Church reform were based on historical considerations of the relationship between Christianity and society. He seemed to be advocating that a thorough knowledge of modern history itself should replace much clerical training in the classics. But even so, it is unclear whether or not Seeley would have considered himself a historian in 1868.

This is because, it is worth recalling, Seeley was trained as a classicist. During his fellowship at Christ's College he tutored and lectured in classics. In 1859 he was the chief classical assistant master at the City of London School. And from 1863 he was professor of Latin at University College, London. For much of his early life, he had worked as a classicist. And, moreover, his published work reflected that fact. In 1869, however, Seeley's career trajectory began to head in a seemingly different direction as he was appointed Regius Professor of Modern

History at Cambridge. "Seemingly" is the key term here, as his new career path was still motivated by his strongly held religious views. With that said, as we will see below, Seeley's publishing strategies did change as he adopted the identity of the historian and sought to avoid the kind of controversy that resulted from the publishing and reception of *Ecce Homo*.

Ironically, his new position was very much owed to *Ecce Homo* and a few of its well-placed supporters. One of those supporters, Charles Kingsley, decided to resign his position as Regius Professor of Modern History at Cambridge in early 1869.[1] Kingsley had been mercilessly attacked by so-called scientific historians such as Edward A. Freeman, who were promoting a disinterested and closely inductive analysis of the development of England's past quite in contrast to the literary approach favoured by Kingsley. To his critics, Kingsley seemed more interested in producing in his students a feeling for the past that had little connection to its documentary traces. Freeman's fellow "Oxford school" historian William Stubbs had, earlier in the decade, been appointed Regius Professor of Modern History at Oxford, where he appeared to be implementing a self-styled Baconian treatment of English history. This would find expression in his much-celebrated *Constitutional History of England*, the first volume of which was published in 1873. There was immense pressure, therefore, for Gladstone to make a comparable appointment in Cambridge. But the position was seen as a fairly thankless one, given the extent of the duties and the relatively poor pay, and his initial offers to Charles Merivale, James Spedding, and Aldis Wright were all rebuffed.[2] Seeley, however, had always been on Gladstone's short list, not just because of his well-documented appreciation of *Ecce Homo* but also because Seeley was both Kingsley's and Maurice's choice for the post.[3]

Kingsley actually did Seeley the favour of writing to him to gauge his interest in the position. Seeley was, of course, flattered at the thought, writing that it was "one of the greatest compliments I have ever received in my life." And he admitted that he was "looking around anxiously" in order to increase his income. The £350 professorial salary would be a £50 improvement from his current position. But he did have significant concerns about his appropriateness for the position. "The truth is," he explained to Kingsley, "that though I have read discursively on modern history and have really given a good deal of thought to philosophies of history I have not studied a simple period of modern history critically in the original authorities." He explained that he had spent the last ten years "polishing my knowledge of Latin and of Ancient history." Tellingly, Seeley said that his decision would be made much easier if Gladstone could be convinced to alter the nature of the position. "Is there any chance of the chair being made one of Universal History? Gladstone, I suppose, can do what he likes with it. If ancient history were included I should not feel so diffident." Not only was Seeley concerned that his training had not prepared him for the position; he was equally concerned about the inevitable "outcry which the Press would raise," a prospect that he admitted "would have some terrors for me."[4] When Gladstone finally made the offer to Seeley, however, it was clearly stressed to him that there would be no title change and that if he were to accept, it would be the study of modern history that he would be expected to teach and promote. Seeley must have put aside his own doubts about his ability and concerns about the inevitable public outcry because he eventually accepted the offer.[5]

As Seeley predicted, however, the press was largely opposed to his appointment, as he seemed to have even less business guiding modern history at Cambridge than did his predecessor. This is what the Master of Trinity, Montagu Butler, was getting at when he found himself surprised that "we should so soon have been regretting poor Kingsley."[6] The fact of the matter was that Seeley simply did not appear to have a track record worthy of the appointment. Making matters worse was that the only relevant book Seeley had published up till then was *Ecce Homo*, which he had yet to acknowledge as his own. His edition of Livy would not appear until 1871. And Gladstone's well-published appreciation of *Ecce Homo* also shed a rather unflattering light on the appointment, as most would not have realized that Seeley was not Gladstone's first choice. The *Saturday Review* was particularly critical of Seeley's appointment in this regard: "Mr. Gladstone was fascinated with Ecce Homo, and therefore Mr. Seeley teaches modern history at Cambridge. He may do it well; but his nomination was quite independent of any sufficiently grounded presumption that he would do so." The *Saturday* continued by explaining how previous office holders at least had a publishing record, whether literary or historical, that placed them in an elite class of writers. Seeley, however, "is not one of this order. He owes his appointment to the personal liking and to the theological sympathy of the Minister of the day. He has still to prove his fitness for the place he fills."[7]

Seeley sought to put his critics' minds at ease by explaining in his inaugural lecture that he would stress the educational value of history, a point that he had made in his earlier essay on "The Church as a Teacher of Morality." Much as in that essay, he stressed that history had to be taught strategically. The past was far too vast a subject to consider in its entirety or randomly. He stressed the need to focus on the history of politics and political institutions and envisioned history becoming the key subject for the training of future statesmen. For Seeley, history was not therefore to be studied for its own sake but for the sake of the present and future. As he said in the last line of his lecture, "If I succeed in any measure, I hope to do so by … giving due precedence in the teaching of History to the present over the past."[8] When we consider these views in light of *Ecce Homo*, it is clear that this was perhaps the driving conception behind that study. He had no interest in characterizing Christ in a particular way and within a particular context for the sake of historical accuracy alone. It was important to represent Christ in that way because it shed light on the contemporary crisis of belief that Seeley witnessed all around him.

Rather than returning to that subject, however, Seeley set about working on just the kind of specialized study that was demanded by the new historiographical standards being promoted, a study that would take him eight years to complete. And, what is more, he also sought to write on a subject that was both undoubtedly modern and political, following precisely from the historical method outlined in his inaugural lecture. The result was a study of Germany during the Napoleonic Age, which focused on the political manoeuvrings of the Prussian statesman Baron vom Stein. As Seeley explained in the preface to *The Life and Times of Stein*, this was not a typical biography but rather an attempt to view an age through the lens of a particularly important historical actor. Seeley was therefore "not influenced by the consideration which so often makes a biographer partial, viz., that if the subject of the biography is not made out to be a very extraordinary man, people will ask why the book was written or why they should read it." Seeley explained that he had no interest in making Stein out to be more interesting than the facts themselves would suggest. "The public are invited to read this book simply because they will find in it an account, the clearest the writer could produce, of a great and momentous transition in the history of Germany."[9] As well as claiming to write a fact-based narrative, unadorned by speculative or dramatic prose, Seeley also gave an account of his sources, which included an impressive set of manuscripts, thereby showing that his account was based entirely on what the primary facts themselves suggested, rather than on secondary authorities. Seeley believed that he had followed a high standard of historical science in producing his three-volume study and he expected that it would be judged according to that standard.[10]

The reception of the book, however, was not quite what Seeley imagined it would be. Or perhaps he was unwilling to see the praise that was often combined with the criticism that tended to focus on the book's unfortunate style, a style that was deemed quite in contrast to the still anonymous *Ecce Homo*. The *Examiner*, for instance, found Seeley's "*début* as an historian" to be highly successful, arguing that the book "inaugurated a new era in English historical method as applied to modern times."[11] In translating the German historical method into English historical practice, the *Examiner* even believed that Seeley had avoided some of the problems that typically came with scientific historical writing, for he "combined very happily the German adhesion to facts and the French devotion to form."[12] It was an "artistic life of Stein, and an admirable description of the times in which he lived and the influence he exerted."[13] The *Saturday Review* similarly found it to be "a valuable contribution to English knowledge of German history and German politics."[14] The reviewer "congratulated [Seeley] on having found a sphere for his labours which thoroughly deserved the patient industry, the exhaustive inquiry, and the discriminating impartiality with which he has approached his task."[15] Embedded in the praise of this review, however, was a criticism that would gain more prominence in future reviews, namely that the book was "wearisome" in places. It was so wearisome in its descriptions of details, argued the *Saturday* reviewer, that "[w]e never know where we are in it. We are swept away into a tide of general reflections and subordinate sketches, and then whirled back into a little eddy of Stein's peculiarities and opinions and anecdotes of his family." This "enormous prolixity" was an unfortunate defect in an otherwise valuable contribution to historical knowledge.[16]

It was precisely the prolixity of the book that many other reviewers could not overlook. And they typically found the book's tedious style much more problematic when assessing its overall value. The *Westminster Review*, for instance, could not get over the length of the book, which it argued was repetitive and filled with "redundant pages."[17] George Strachey for the *Academy* argued that what the book needed was less of a summary of facts and more "systematic grouping and crystallisation." Strachey said that "Prof. Seeley tells rather too much of his story with the scissors, and not quite enough with the pen."[18] This was also the view of the *Spectator*: "[Seeley] has thought it necessary to give all the facts and many of the documents upon which his opinions are founded, to print all he knows of his hero's official work, and to tell us all the circumstances by which he found himself impeded, and the consequence is that we often lost sight of the man himself, without gaining any idea of general history."

This was unfortunate, according to the *Spectator*, because it meant that Seeley's inability to organize and condense his material undermined the impressive research that went into the three volumes. And this was ultimately a problem with the author's "style." "A deficiency almost complete in the power of pictorial writing has spoiled the result of great labour, great knowledge, and much careful thought." The book is so full of information that it would be perfect for a historian looking to find materials to form the basis for a sketch about Stein, but as a biography itself, it is simply a failure, as "it wearies the reader who sought for a ... history, and only finds annals, and who longs in vain for a stirring chapter, or a memorable passage, or even a brilliant paragraph. There is none such throughout the work. Instead of these, he finds an even but sometimes dull flow of very minute narrative, very impartial, very accurate, and occasionally ... very suggestive, but on the whole uninteresting, and even tedious. The materials have not been thoroughly worked up, and the result is dough, not bread."[19] The *International Review* was particularly harsh in its suggestion that "the volumes are hard reading. There is so much analysis and discussion that the reader is never caught by the story and swept along ... by the great events of time. This is a grave defect, both for author and public."[20] The *London Society* similarly found the book's literary merits wanting, believing that it was far lesser in this regard than the popular histories of the day. "Compared with such a writer as [Thomas Babington] Macaulay," argued the *London Society*, "Professor Seeley is dull."[21]

Despite the fact that just about all reviewers found Seeley's scholarship at the front rank of English historians, he was clearly hurt by the negative commentary about the style of his writing, which was deemed responsible for the book's failure to reach anything but a specialized audience. In his correspondence with Robert Spence Watson, who was an admirer of the book, Seeley expressed astonishment at how the *Life and Times of Stein* was received by his supposed peers. "It gave me great delight to read your letter on Stein and I feel heartily obliged to you for taking the trouble to write down your thoughts. It has been a difficult hill to climb and it is of course trying to me to find many of my friends, who have long been telling me with what eagerness they have expected the book, now silent, and embarrassed and betraying plainly that they cannot read it. To some extent I was prepared for this and at any rate I remain firmly convinced that I have hit on the right way of writing history and that the Macaulay style is radically wrong. But I should have been sadly vexed to hear nothing but the refrain 'Very solid, very valuable, but dreadfully dull.' I wanted to feel that upon those who could read the book through it was capable of producing a powerful effect, and so your letter did a great deal towards keeping my heart up."[22] He later wrote to Watson that he recognized that the "book was somewhat hard reading in parts, that it was by no means like a romance, but made no doubt that scholars and educated men would stand by me and would tell the public that my book was all the better on that account and that the notion that a history ought to be a delightful Mudie book was childish." It was therefore surprising to him that his peers seemed to have "the same opinion as the vulgar." This was perhaps most frustrating for Seeley because it suggested that his previous readers now believed that he had "grown stupid" while his friends treated the book "as an awkward blunder that I have committed." The reception of the book was the sign of a larger problem for Seeley and of the great extent to which "the public taste" for history had become "demoralised." "Is it not a very deplorable and contemptible condition of the English mind that it should condemn a book avowedly for nothing but its thoroughness and a style for nothing but its severity? I notice that people have lost the very idea of a severe style and think that I wished to be what they call 'brilliant' but could not do it."[23]

The reception of his *Life and Times of Stein* inspired Seeley to write a series of articles on "History and Politics" for *Macmillan's Magazine* that set out to justify his own "style" of writing history in contrast to that favoured by the "demoralised" general public, the "Macaulay style" and all that entailed.[24] In a way, Seeley was following other so-called scientific historians of the period who challenged the previous generation's novelistic approach that tended to promote readability at the expense of accurately representing the facts of the past. But whereas someone such as Seeley's future counterpart at Oxford, Edward A. Freeman (who would replace William Stubbs in 1884), argued that the general public needed to be re-educated to appreciate the new scientific standards that might be less palatable to their imaginations, Seeley argued that these readers were a lost cause, and that historians should essentially write for peers alone. Moreover, Seeley argued that it was largely Carlyle and Macaulay who were to be blamed for corrupting the general reader's historical sensibilities. This was not exactly a controversial claim in regard to Carlyle, but Macaulay had largely been immune from such criticism. For Freeman, for instance, Macaulay more or less combined the artistic and empirical sides of historical narrative that the new scientific historian ought to emulate. Seeley, however, clearly disagreed.[25]

Macaulay's histories were to be particularly lamented, argued Seeley, because they appeared to be based on empirical research but were in fact more interested in dramatizing the past to please general readers brought up on the romantic historical fiction of Walter Scott or one of his many imitators. Macaulay, along with Carlyle, had spoiled the public taste for factual history, making readers think that the past is itself inherently dramatic and interesting, so that they fail to appreciate a history that represents the past as it actually happened. As Seeley explained to C.E. Maurice, "I urge against Macaulay and Carlyle that they sinned against light and so betrayed minds radically unscientific … To my mind these two men may be expected to be remembered some day as representing an extraordinary aberration in the English mind, an extraordinary misconception of the nature of history. This is my opinion which evidently is so surprising to you that you cannot even imagine me to hold it."[26]

Maurice was likely surprised because this opinion did seem to contradict views previously expressed by Seeley. When he was a young man, around the time he was the chief classical assistant master at the City of London School, Seeley attempted to persuade his sister Mary to "persevere a little in trying to read Carlyle's Friedrich," by which he meant Carlyle's *History of Friedrich II of Prussia* (1858). Even then Seeley appreciated that much of what Carlyle said was factually inaccurate, but he told Mary that it was more important to realize that it "has a truth in it." And perhaps most relevantly, he believed that Carlyle was right to criticize most other historians for their inability to make history interesting: "Besides, the ordinary historians, the Dryasdusts, really deserve some of his censure with their solemn judicial decisions and platitudes."[27] The Seeley writing these words was very different from the Seeley who was writing platitudes in *Macmillan's Magazine* about the dignity of scientific history.

We certainly should not be surprised that Seeley changed his views during the space of twenty years. What is surprising, however, is that his views seemed to change so diametrically. In this regard J. Llewelyn Davies was once again suggestive in his observations, arguing that he found it quite "difficult to understand how the author of Ecce Homo came to write in the manner of the Life of Stein." For his part, Davies took Seeley at his word that he was "so repelled by the chorus of Macaulay–Froude that he resolved to try how perfectly he could strip a history of all that an ordinary reader enjoys."[28] This brings up an important question, however, about just what was the relationship between *Ecce Homo* and *The Life and Times of Stein*, not only in terms of their different styles but also in terms of what they suggest abut Seeley's evolving thoughts about the use of history more generally.

It is true that the two works were viewed at the time as being written for entirely different purposes, for different audiences, and underpinned by different methods. But Seeley did not see them that way. Indeed, for him the two books were intimately connected. As we have seen, *Ecce Homo* was a study of the ancient world explicitly through the perspective of the present, with its ultimate message being that the teachings of Christ could provide the moral foundation for a more integrated relationship between political society and religion. For Seeley, Stein represented just this approach in his shaping of the modern German nation.

It is telling that Seeley believed that the best way to write a book about a nation in transition was to focus on an individual. "When I first asked myself how this subject might best be treated, I remarked that it was one which especially needed to be gathered up in a person."[29] Moreover, while Seeley had no clear idea at the time about how Stein had been represented in previous studies, he came to believe that he was partially, if not entirely, misunderstood. It then became Seeley's goal to present the character of Stein as accurately as possible, free of the mistakes of "careless second-hand sketch[es] ... When I was fairly embarked I soon found the character much more interesting than I had anticipated, and before long it had become a labour of love to me to shed the clearest possible light upon it."[30] As Duncan Bell has argued, Seeley regarded Stein "as a founding 'father' of modern Germany, and as portending many of the crucial developments of the ensuing century."[31] There is a clear analogy at work here between the way he represented Stein and the way he represented Christ as the founder of a moral society capable of adapting to changing political and social circumstances. Both men were called upon during moments of immense transition and both established political entities that could progress along with history itself.

While Seeley rarely reflected on the relationship between the two books, there is an oft-recounted episode that gives some evidence to suggest that he viewed them as parts of the same larger project. This is at least anecdotally apparent in Caroline Jebb's biography of her husband, Richard Claverhouse Jebb, the classical scholar, who was a friend of Seeley's. There, Caroline Jebb recounts a response Seeley gave to her husband's question about whether Seeley ever planned on writing the sequel to *Ecce Homo*. "The answer, most unexpected, was to the effect that he had fulfilled this intention already. On being pressed for an explanation, he said that he meant his Life of Stein! His questioner's comment on this ... was that if he had heard this statement attributed to Seeley, he would have scouted it as incredible."[32] The real irony in this story, however, is that Seeley *had* already by this time written the sequel to *Ecce Homo* and it was not published as *The Life of and Times of Stein*. The reason Jebb did not know about it was because of the way it was published, which is the subject of the next chapter.

Chapter Twelve

Fulfilling a Promise

The success of Ecce Home was rather alarming than otherwise. If I knew any way in which to prevent all weak or rash heads from reading it, I would certainly adopt it.

– Seeley to Macmillan, 7 July 1872

At the same time as Seeley was seeking to shape the scientific study of modern history in Britain, and while he was writing his three-volume study of Germany and Stein, he had not completely forgotten about his previous life as an anonymous author of a sensational study of the life of Jesus Christ. Macmillan remained in contact with Seeley about his various projects, so any bitterness about the handling of *Ecce Homo*'s authorship must have been short-lived. However, an important qualification must be added to this statement. Seeley was now much less willing to negotiate how his future work would be published, whether it would suit Macmillan's bottom line or not. And Macmillan was seemingly happy to publish whatever Seeley sent his way, whether it was an article denouncing Macaulay for the magazine or a collection of Cambridge lectures.[1] But even while Macmillan was cultivating an important network of historians whom he could depend on to produce a variety of timely histories, a network that included Seeley,[2] works of a theological interest remained an important part of the Macmillan catalogue. Macmillan therefore often inquired about Seeley's theological writings, clearly hoping that Seeley had not forgotten that *Ecce Homo* was originally conceived as the first of two volumes. It was, after all, often advertised that way. The Roberts Brothers edition published out of Boston, for instance, included a message on the verso of the half-title, facing the title page, which read: "In Preparation: *Christ as the Creator of*

Modern Theology and Religion. By the Author of 'Ecce Homo.'" The statement was signed "Roberts Brothers, Publishers, Boston."[3]

Many readers of *Ecce Homo*, therefore, wanted to hold off their ultimate judgment until the second volume was published, and that was because much of the justification for the methodology and fragmentary nature of *Ecce Homo* was made precisely under the assumption that a second volume would unite Jesus the man with Jesus the saviour while clarifying the book's many ambiguities. No doubt that was why the Roberts Brothers edition included the statement that the future volume was supposedly "in preparation." The *Dublin University Magazine*, for instance, was just one of many reviewers of *Ecce Homo* that pointed out "that the whole theory is shrouded in a mantle of uncertainty by a promise of another work, which will go more deeply into the theological part of the question. It is not improbable that this reserve to which the author occasionally retreats in moments of danger may contain modifications of what is advanced here."[4] Some readers grew so impatient waiting for the second volume that they produced it themselves. Gladstone's *On Ecce Homo*, for instance, was largely written in an attempt to explore those theological grounds Seeley had left untouched. Perhaps more relevant was an anonymous volume entitled *Ecce Deus: Essays on the Life and Doctrine of Jesus Christ with Controversial Notes on "Ecce Homo."* This book was written in direct response to *Ecce Homo* precisely to bring together the story of Christ the "Son of Man" with Christ the "Son of God."[5] The manuscript was originally sent to Macmillan in June 1866, and apparently the unknown author wanted it to be "printed uniform with Ecce Homo." Macmillan wrote to Seeley about the manuscript and suggested that the writing was even "a clever kind of aping of your style throughout." That said, Macmillan was adamant that he "would not dream of taking the book on any terms."[6] It was eventually published the following year with the Edinburgh publishers T & T Clark. Macmillan would therefore not let Seeley forget that much of the reception of *Ecce Homo* was based on this promise that had gone unfulfilled. That it had gone unfulfilled was certainly understandable, as Seeley was beginning his new and very demanding position as Regius Professor of Modern History at Cambridge, which did not leave much time for work that was not directly relevant to his study of Stein or his lectures on modern history. He was therefore happy for the time being to let others speculate about what that future study might look like.

By 1872, however, Macmillan began inquiring if Seeley had been lately able to write on theological topics, and Seeley's response proved to be about as optimistic as Macmillan could have hoped for. "You asked to see what I have been doing in theology," Seeley responded on 25 June 1872. "I told you it was too fragmentary to be publishable, but on reflection I should be glad if you would look through what I have written ... You will remember that it is in a very rough state and also that it is a very small part of what I have in my head."[7] After reading the still unfinished manuscript, Macmillan responded enthusiastically that "it is as regards mere style and treatment quite equal to Ecce Homo." Macmillan did recognize that the "*interest*" would likely be less intense because of the "abstract nature of the subjects you deal with." At the same time, he expected that, whatever was to become of the "unfinished work," it would be "a not inconsequential success."[8] And, eventually, these "fragmentary" thoughts on theology would end up forming the basis of a series of ten articles that were published in *Macmillan's Magazine*, articles that would later be revised and published in book form as *Natural Religion* (1882), the long-promised sequel to *Ecce Homo*.[9]

From the first, however, it became clear that Macmillan and Seeley had a different understanding about how these original fragmentary thoughts would be written, published, and then marketed. This is apparent in Seeley's response to Macmillan quoted above. Seeley thanked Macmillan for his "encouraging ... notes," which pointed to "some inconsistencies and contradictions." About the fact that the new study would likely interest readers less than did *Ecce Homo*, Seeley suggested that producing a less popular work was actually his intention. "Of course one would not like the book to be a failure," Seeley expanded, "but beyond that I really do not care much. The success of Ecce Homo was rather alarming than otherwise. If I knew any way in which to prevent all weak or rash heads from reading it, I would certainly adopt it. It is this feeling which leads me to delay so much, I mean the feeling that however much good I may hope to do I cannot fail also to do harm."[10] This is, I think, an important comment, giving us much insight into how Seeley's post–*Ecce Homo* publications – particularly but not only *Natural Religion* – were written with a view to avoiding precisely the kind of controversy that was engendered by *Ecce Homo*. He wanted to prevent "weak or rash heads from reading" his work because the success of *Ecce Homo* brought not just pleasure but also the pain of being confronted by family members and other leading members of society who found his work dangerous. The problem as Macmillan saw it, however, was that in order to ensure avoiding the controversy that had been associated with *Ecce Homo*'s success, Seeley was going to produce a book that would be a failure, a concern that Seeley did not share.

Perhaps most problematic from Macmillan's point of view, however, was that Seeley claimed that he wanted the articles published anonymously, without a signature of any kind. This was awkward, particularly for *Macmillan's Magazine*. While anonymity had been largely the rule of journalism before mid-century, as authors of periodical articles were expected to adopt the particular editorial line of a given proprietor, this began to change in the second half of the century: many new journals expected their contributors to sign their names and stand by their opinions, not as being promoted by the proprietors but rather as produced by the individuals writing the articles. When *Macmillan's Magazine* appeared in 1859, the first issue included seven articles, six of which were signed by their authors (the other was signed by "the author of *Tom Brown's School Days*").[11] This was significant because it suggested that journals could feature widely ranging opinions on a variety of topics that need not follow the directives of a particular party or corporate entity. And this happened to conform nicely to the burgeoning liberal ideology that fetishized the "individual opinion," an ideology that was central to many of the new journals that followed the practice of *Macmillan's Magazine*. The very same year that *Ecce Homo* was first published, for instance, the *Fortnightly Review* was established under the explicit editorial policy that the new journal's articles would be signed by their authors.[12]

Macmillan's Magazine never made signature an absolute editorial policy per se, but it is clear that Macmillan and the journal's first editor, David Masson, preferred to publish articles that included a signature. When Thomas Hughes, for instance, wrote a piece for *Macmillan's Magazine* on the Paris World's Fair in 1867, Macmillan was adamant that Hughes sign his name.[13] On 21 October he wrote to Masson that "Hughes's Paris paper ought to be an excellent addition. We must make him give his name, unless he has strong reasons and won't."[14] A day later Macmillan re-emphasized that "unless Hughes has given strong instructions to the contrary, we *must* have his name."[15] Interestingly, the same issue in which Hughes's essay appears under his own name also includes a printed lecture on "English in Schools," signed by "Professor Seeley, MA."[16] And just as Macmillan sought to get Hughes to sign, he did the same with Seeley: "I hope we have Seeley's name," he wrote to Masson: "it will be good."[17]

While it was certainly beneficial to the magazine's bottom line to have the articles signed by well-known public figures such as Seeley and Hughes, the argument in favour of the signature was being powerfully made for non-commercial reasons, particularly by just the kind of liberal figures that *Macmillan's Magazine* was associated with. Hughes himself, for instance, wrote a piece in the magazine that was heavily critical of anonymous publishing. It was ostensibly written as a critique of an essay in *The Times* that seemed to take for granted that anonymity was equated with good and proper journalism. Hughes challenged that assumption. And as someone who had previously written anonymously, he wrote his piece as if he had now come to see the errors of his ways. He argued, fundamentally, that the anonymous author holds an insidious power over his readers that makes them think they are reading an opinion "representative of some great unknown which haunts the majority of readers of newspapers" rather than one simply produced by a single person who happens to be "shrouded in mystery." The sort of power that a man gains from such a situation "is not genuine, and can benefit neither himself nor anyone else."[18]

The fact remained for Hughes that anonymous writing really only benefited three classes of people: "First, the proprietors, whose property is made more valuable by the custom. Secondly, the editors, who gain importance and prestige from the sort of mystery in which they are able to wrap themselves. Thirdly, we the writers, who, while the custom prevails, can write with much less a sense of responsibility, and therefore much more copiously and easily." Notably, the three classes that benefit do not include any readers, which of course was Hughes's point. But given that these three classes were the ones essentially controlling journalism, the practice of anonymity was one that was difficult to challenge. As Hughes argued, "The interest of all these three classes lies in the same direction, that of prolonging the reign of the mighty 'we.'"[19] The royal "we," therefore, represented an older and outdated form of journalism that it was now time to abandon.

Hughes also made it clear that he was not alone in adhering to this view, that it was shared by many of his friends and colleagues: "Besides, to judge from my own experience, so far from there being no doubt that anonymous writing is the only eligible and effective form of newspaper writing, I find the persons amongst whom I live constantly debating the point whether anonymous writing ought to be tolerated."[20] This was no doubt true of many of the figures associated with Macmillan and the magazine. F.D. Maurice, for instance, who as we have seen was an important contributor to Macmillan's catalogue and was also a central figure in the early history of the magazine,[21] was an outspoken critic of anonymous journalism. When he was advising Macmillan and others interested in founding the new journal, his central concern was that the journal should only include signed articles. He wrote to Charles Kingsley that "About the general maxim that it is desirable, nay, absolutely necessary to begin, and to begin speedily, some journal which shall discard anonymous writing, I have no doubt whatever."[22] He argued, much as Hughes would later, that "the *We*" that represents anonymity "covers the most insolent pretension, the most offhand dogmatism, the most haughty scorn of individuals and of mankind." The time had now come, he argued, "when those whose circumstances or whose ignorance disqualify them for those higher tasks, but who feel that there is a foul stench sent forth by our anonymous periodical literature and that they in times past if not now have contributed, as I have, to the increase of it, may think that they are bound before they die to do something, be it ever so little, for the purpose of purifying the moral atmosphere of the vanity, cruelty, falsehood with which it is impregnated."[23] A new journal, relying on the labour of those willing to stand by their public opinions and sign their names, who "dare not be braggarts and dare not be cowards," are thereby "arming themselves, and so far as they can are assisting others to arm, for a real battle, in which impertinence and frivolity cannot serve except to expose those who indulge in them."[24]

In a subsequent letter to R.H. Hutton, who had previously written to Maurice in defence of anonymous journalism, Maurice suggested that what was really at issue was that, as currently practised, anonymity promoted a harmful sectarianism. "Now it seems to me that journalism, with its symbol We, has more than anything else tended to keep alive the notion of sect fellowship, by which I mean fellowship in a certain more or less accurately defined set of opinions." The fact that so many journals subscribed to particular religious doctrines and thereby expected their authors to promote those positions meant that actual engagement with the problems of the day was being ignored in favour of bringing out "that which was most contentious and negative in the persons whom they represented."[25] Because anonymity encouraged writers to stand not behind their own personal views but rather behind those of the proprietor, it could only contribute to the maintenance of competing corporations and sects. By attempting to get rid of anonymous journalism, Maurice believed that more individuals would seek to transcend those divisions rather than maintain them. This was ultimately what motivated Maurice's position: "Because I desire that all men should regard themselves as forming one fellowship in a real and living Head," Maurice continued, "the formation of sects and corporations, grounded upon mere similarity of opinions, is that which I most dread."[26]

Seeley would have agreed with the premises of Maurice's attack on anonymous journalism. He, too, believed in the universalism of Christianity. As we have seen, one of the main themes of *Ecce Homo* was the way in which Christ's morality developed out of an enthusiasm of humanity, an enthusiasm based on the fact that all humans are God's children. Moreover, Seeley wrote *Ecce Homo* in the hopes that he could convince Christians of all denominations as well as growing numbers of unbelievers to embrace an interpretation of Christ that was free of the doctrinal and theological conceptions that had led to so much division and scepticism. We have also seen, however, that this was a message that could only be conveyed anonymously. It was only once conceptions of Christ were not attached to particular names, and therefore particular party identities, that a new characterization of Christ could emerge. With that said, however, anonymous book publishing was not the same as anonymous journalism. Anonymous authors of books were not necessarily seen as shadow versions of their publishers and editors in the same way that anonymous journalists were seen as representing the corporate views of their proprietors. Seeley would have therefore recognized that *Ecce Homo*'s methodological rationale for anonymity could not similarly be made for its sequel if it was to be serialized in a periodical. If publishing *Ecce Homo* anonymously divided some of its supporters, the decision to publish its sequel anonymously in *Macmillan's Magazine* went entirely against the growing consensus against anonymous journalism among Seeley's liberal friends.

It is clear that Seeley was not adopting some royal "we" to give his work an extra shadowy authority or conform his views to a particular sect of opinions. Indeed, it seems that his appropriation of the tool of anonymity was of a very different order. If there was a defence of anonymous journalism at the time that comes close to grasping Seeley's purpose, it is one that appeared, anonymously of course, in *Blackwood's Magazine*. The author, the Scottish journalist and critic E.S. Dallas, argued that by eschewing a signature, the author "must write on public grounds, it is no longer Smith, who is compelled by his invisible cap to forget that part of his nature which is peculiar to himself and essentially private – Smith, who is forced to regard only that part of his consciousness which identifies him with every other member of the community – Smith, no longer the individual unit, but the representative man."[27]

At least at this point, Seeley did not want his name to bring undue attention to the essays. Moreover, as with *Ecce Homo*, he was seeking to appropriate a particular public voice in order to follow a particular line of reasoning for the sake of argument. He always believed that any name that was necessarily attached to such a project would necessarily skew its reception. It was ironic that in writing anonymously, Seeley was attempting to avoid being pigeonholed as representing a particular party line, when anonymous journalism was associated precisely with the kind of party writing that he was trying to avoid.

It is clear that Macmillan and the editors at *Macmillan's Magazine* would have preferred Seeley to sign his name to the articles. In this instance, Hughes was decidedly wrong in his analysis that anonymous writing suited the proprietors and the editors. Editorial policy of signature aside, Macmillan doubted that, left unsigned, the articles would receive much notice at all. So while the debate about anonymity vs signature lay just below the surface of Macmillan and Seeley's debate about the appropriate publishing of the "Natural Religion" essays, ultimately Macmillan was concerned about securing a readership for the articles. In this regard, according to Macmillan, anonymity suited no one. Macmillan agreed to relent, at least initially, accepting Seeley's suggestion that they should "at any rate try the experiment of anonymous and see how it works."[28] By all accounts, it did not work well.

Throughout 1875 the first five instalments of "Natural Religion" appeared in *Macmillan's Magazine*, and in the articles Seeley set out to consider the various similarities between scientific and Christian ways of thinking in order to bring about a reconciliation between the two seemingly conflictual modes of thought while establishing a form of Christianity that would be conducive to naturalistic conceptions of life. The articles, therefore, offered a kind of "natural" religion to replace the current dominant conception of Christianity as a "supernatural" religion. In hindsight, the articles certainly shed light on the theology that underpinned *Ecce Homo*, and we will have course to discuss this more thoroughly in the next chapter, but it would have been unclear to readers that the articles were in any way connected to *Ecce Homo*, given that the original study is never mentioned, nor is there any suggestion that the articles were following a previous publication. Moreover, Christ is not the central figure of the essays, which readers of *Ecce Homo* would certainly have expected of a sequel. Seeley originally said himself that a sequel would deal explicitly with Christ as "creator of modern theology and religion," but this was not the central focus of the essays. In this regard, the *Dublin University Magazine* was right to be concerned that any sequel "may contain modifications of what is advanced" in *Ecce Homo*.[29] Readers could therefore not be expected to seek out the various hints and clues as to the articles' ultimate origins in the same way that they sought to do in order to discover the identity of *Ecce Homo*'s hidden author.

It would have been almost impossible, therefore, for readers to engage in the kind of deep readings that would have been required to acknowledge the connection. There was very little in the articles that explicitly connected this study with its much more popular predecessor. So by the end of the year it was clear that the essays were being largely ignored. Macmillan seems to have wanted the future essays to make the connection to *Ecce Homo* more apparent, perhaps by signing them "by the author of *Ecce Homo*." Seeley was unwilling to accept such a proposal, however, as he wanted to remain well and truly anonymous. Connecting the articles in any way to *Ecce Homo* would undermine that status, given that he was already well known as its reputed author. He wrote to George Grove, who replaced Masson as editor of *Macmillan's Magazine*, at the end of 1875 with just that message:

> I quite agree with you and Macmillan that these papers may possibly attract no attention, nay I should say even probably if there were only one of them. It may seem to you whimsical that I should choose to run the risk of such a failure when I have a ready means of securing attention. It would be very whimsical if the question were of a poem or a novel, but on more burning subjects popularity brings much more pain than pleasure, and besides pain perplexity and anxiety. The truth is I had much rather fail as you anticipate than succeed in the other way. But I think I have a fair chance of getting the kind of attention I want. A series of articles extending over several months and pressing one point home have a chance of being attended to that a single article has not, particularly as your magazine goes into so many hands. You kindly allow me to put out of sight the interests of your magazine. At the same time do not let me ask you to burden yourself with a number of anonymous papers which you do not care about – but I thought that in these theological days you would not object to ... some theological articles which I hope have at any rate some flourishes of style.[30]

Seeley recognized that these essays, because they were anonymous, were somewhat of a "burden" to the magazine. They were a burden to its editorial policy of signature. And Seeley's name, along with the connection the articles had to *Ecce Homo*, was indeed "a ready means of securing attention." Had the articles been written in a much more attractive style, perhaps they would not have required Seeley's name. As it was, however, Seeley seemed to write the articles in such a way that they could not possibly secure much of an audience. And this was apparently by design and followed from Seeley's newly adopted persona as a serious and scientific historian. These concerns would also play a role in Seeley's considerations of publishing the articles in book form.

Even though Grove and Macmillan repeatedly pressured Seeley into avoiding long intervals between the article instalments of his "Natural Religion" series,[31] the gaps between the articles grew longer and longer as the series progressed. And as the series appeared to lurch towards an end in 1877 when the ninth article appeared, it became clear that the series failed to find much of a readership. On 11 January 1877, Macmillan wrote to Seeley, lamenting about how the articles were published, giving us insight into what he would have preferred: "I wish you had allowed us to put by the Author of — to the articles in the Magazine. It would have done good all around."[32] There was even a suggestion made that Seeley should sign the last article "by the author of *Ecce Homo*," but that was rebuffed as well.[33] "I am sorry to feel obliged to keep back my name," Seeley wrote to Macmillan, "particularly as you say it would help you to publish it. But I told Grove how I felt about it at the beginning."[34]

Undaunted, Macmillan was keen to explore the possibility of republishing the essays in book form, as there was only one article left to publish: "Might we not arrange about the reprint in book form of Natural Religion?" Macmillan wrote to Seeley at the beginning of 1877. "When can you come over and talk about it?"[35] Seeley was no doubt surprised that Macmillan still wanted to pursue a book, given the lack of interest in the series itself, responding that he was "obliged" to Macmillan "for being the first to mention reprinting in book-form articles which have been so much neglected." But he said that he was several months away from finishing the final article in the series, and was otherwise far too busy to concern himself with a reprint. "The fact is I feel as if I was writing three books at once," Seeley wrote to Macmillan on 6 February 1877, "of which two require a great deal of research and the third a great deal of thought." He was, most importantly, still writing *The Life and Times of Stein*, and was determined to finish it by the end of July. "Meanwhile I am beginning two courses of lectures on English History, which taken together cost me quite as much trouble as would make a book. Nat[ural] Rel[igion] is the third." Seeley held out hope, however, that the final instalment could be finished soon, at which point he could think about the book: "it may be possible by breaking the Sabbath three or four times to get the concluding article written in time for your June number [of *Macmillan's Magazine*], and then I shall be ready to think of arranging a reprint."[36]

As if on cue, Macmillan wrote to Seeley in June. Was he "ready to discuss the republication of 'Natural Religion?'" Macmillan had "little doubt," he told Seeley, "about its success in a completed form."[37] The final article, however, was still not yet completed. The situation would not be clarified until it was finally finished and set in type in December 1877, just in time for the January 1878 issue. Macmillan then proposed to send Seeley a copy of all ten of the papers "interleaved" so that Seeley could simply insert any changes directly. He was keen that the book be published "as soon as possible after" the last instalment of the series.[38]

But Seeley again hesitated. The problem now was that *The Life and Times of Stein* was finished and was scheduled to appear in 1878, and would therefore generally overlap with the time of publication Macmillan was proposing for *Natural Religion*. Seeley did not want readers to confuse the two works. *The Life and Times of Stein* was meant to establish Seeley's credentials as a scientific historian, a reputation that, as we have seen, he had been working carefully to build since his professorial appointment in 1869. He was worried that all that hard work would be undone. Macmillan recognized Seeley's predicament but offered a telling solution: "all you say has weight," Macmillan wrote to Seeley, "but if you publish it as by the author of 'Ecce Homo' and not the Professor a good deal of the objection to immediate publication would vanish." Macmillan argued that "readers of Stein would read the other [book] in many cases without a distinct consciousness that it was by the same author." Readers would, in other words, recognize that they were meant to read the two books differently, as signified by the particular signature.[39]

While Seeley remained unconvinced about publishing *The Life and Times of Stein* and *Natural Religion* concurrently, he does seem to have been convinced by Macmillan's logic for including a signature on the title page of *Natural Religion*, suddenly reversing his previous commitment to absolute anonymity. He did continue to delay the publication of *Natural Religion*, however, as he debated revising the articles more thoroughly, and then was determined to first publish his series of articles on "History and Politics," articles that as we have seen (chapter 11) acted as a response to the reception of *The Life and Times of Stein*.[40] Macmillan would not give up. He kept asking Seeley when the press could expect the revised book.[41] And finally, on 28 January 1880, Macmillan had his answer: "I am glad you mention Natural Religion," Seeley wrote to Macmillan, as if it was the first time Macmillan had broached the topic. "I had been thinking about it a good deal lately … My original notion was to alter a good deal before publishing it, but now think I may leave it much as it was, and only add a preface."[42] Macmillan was delighted and immediately arranged for the articles to be set in type.[43] Seeley then spent the better part of a year writing the preface, and received more pleas from Macmillan to speed the process along.[44] *Natural Religion* was finally published, four years after the conclusion of the series, in July 1882. And this time there was no doubt about its authorship. It was signed "by the Author of *Ecce Homo*" (see figure 13.1).

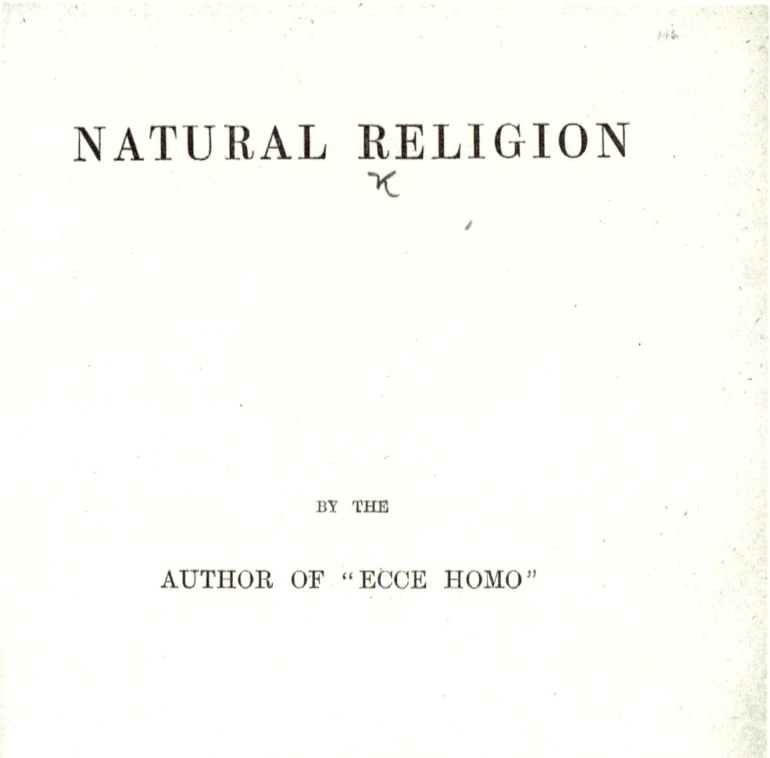

NATURAL RELIGION

BY THE

AUTHOR OF "ECCE HOMO"

"We live by admiration."
WORDSWORTH.

London:

MACMILLAN AND CO.

1882

The Right of Translation and Reproduction is reserved.

Figure 13.1 *Natural Religion, by the Author of "Ecce Homo"* (London: Macmillan, 1882). © The British Library Board. General Reference Collection 4109.k.8.

Chapter Thirteen

By the Author of *Ecce Homo*

By reviving prophecy in its modern form of a philosophy of history, we at once adapt religion to the present age and restore it to its original character.
– *Natural Religion, by the Author of "Ecce Homo"* (1882)

Instead of the bread we hoped for, a stone has been thrown to us; instead of a fish we have been mocked with a serpent.
– "Natural Religion," *Quarterly Review* 154 (October 1882)

Macmillan had high hopes for the new book, he wrote to Seeley, "especially when it is clear that it is by the author of 'Ecce Homo.'"[1] It is not entirely clear just how Macmillan convinced Seeley to drop his initial demand of absolute anonymity. But he most certainly did. And by agreeing to sign the book "by the Author of *Ecce Homo*," Seeley was now following a well-worn publishing practice, one that was also followed by his father (see chapter 2), essentially advertising the new book with the title of a previously successful one. Moreover, as Macmillan made clear to Seeley (see chapter 12), by signing the new book in this way, they were encouraging readers to remember the experience they had when they read *Ecce Homo* under the assumption that just such an experience was now awaiting them when reading the new study. *Natural Religion* was, therefore, meant to be understood as part of a bibliographic series that began with *Ecce Homo*. However, aside from the signature, the only explicit mention of *Ecce Homo* appears not in the main text but in the paratext, namely as an advertisement at the end of the book under the heading "*By the same author.*"[2]

Natural Religion did include, however, a short two-page preface, like *Ecce Homo*, and it proved to be just as ambiguous as the preface in the original. After stating that the book originated as a series of articles in *Macmillan's Magazine*, Seeley pointed out that about two-thirds of the book remained the same as the original articles, while "the remaining third part is wholly new in substance as well as in style."[3] He went on explicitly to adopt the identity that was promoted about him by the most favourable reviews of *Ecce Homo*, that he was not a representative of any Church party but rather "one of those simpletons who believe that, alike in politics and religion, there are truths outside the region of party debate, and that these truths are more important than the contending parties will easily be induced to believe."[4] Readers "with the expectation of finding … anything calculated to promote either orthodoxy or heterodoxy" were therefore cautioned to seek out such work elsewhere.[5]

Seeley gives no sense about what might follow in terms of content or argument but does instead make clear his view that he interprets religion and politics as "almost one and the same," which certainly follows from the main narrative of *Ecce Homo*. In this regard, Seeley presented himself as standing above the fray of the hopeless debates that informed both of these realms as "he watches with a kind of despair the infatuation of party-spirit gradually surrendering the whole area to dispute and denial and despising as insipid whatever is not controvertible, until perhaps at last, when the brawl subsides from mere exhaustion, a third party is heard proclaiming that when eleven men differ so much and so long, it is evidence that nothing can be known, and possible even that there is nothing to know." His short preface concluded with a rather depressing thought about what might happen to a nation so caught up in divisive debates, and here he alludes to "the catastrophe of Poland, which found such a fatal enjoyment in quarrelling and quarrelled so long, that a day came at last when there was no Poland any more, and then quarrelling ceased."[6]

The stakes had clearly grown since the publication of *Ecce Homo*. If Seeley was in 1865 concerned about the role a burgeoning science was playing in leading Christians towards a vague form of unbelief, by 1882 he believed that the situation had only worsened. According to Seeley, the antagonism between science and religion had grown in recent years while the division between them had widened. Seeley was not, however, an advocate of the so-called conflict thesis that was being perpetuated in intellectual histories such as John William Draper's *History of the Conflict between Religion and Science* (1874).[7] Quite the contrary. What Seeley wanted to show was that religion and science, when properly understood, were entirely, indeed necessarily, complementary. He argued that a modern religion needed to embrace a scientific perspective of life and that a modern science needed to embrace a moral framework that could only be derived from religion. Seeley therefore hoped that *Natural Religion* would also act as a map of the future that would not only help preserve a form of Christianity but along with it the English nation and modern civilization itself.

It has been argued that *Natural Religion* was probably not intended to be the long-promised sequel to *Ecce Homo*.[8] And, indeed, if we were to go by Seeley's statement in *Ecce Homo*'s preface, that a sequel would deal with "Christ as the Creator of Modern Theology and Religion," we would surely have to agree. But there are other moments in *Ecce Homo*, which gesture towards the future volume, that anticipate what ultimately became *Natural Religion* much more clearly. In the conclusion of *Ecce Homo*, for instance, Seeley argued that the present volume's "object has hitherto been Christian morality," and he was particularly concerned with showing "the scheme by which Christ united men together" under a universal conception of mankind. In this way, Christ enabled man to overcome an important enemy: himself. "But," Seeley stressed, "man has other enemies beside himself, and has need of protections and supports which morality cannot give. He is at enmity with Nature as well as with his brother-man." The problem with nature as an enemy in contrast to man is that it is possible to understand the harm caused by man. "But when the forces of Nature become hostile to us, we know neither why it is so nor what to do."[9] What is particularly difficult to understand about nature, Seeley argued, is the mystery known as death. It is "death," according to Seeley, that "remains the fatal bar to all complete satisfaction, the disturber of all great plans, the Nemesis of all great happiness, the standing dire discouragement of human nature … What comfort Christ gave men under these evils, how he reconciled them to nature as well as to each other by offering to them new views of the Power by which the world is governed, by his own triumph over death, and by his revelation of eternity, will be the subject of another treatise."[10]

While it is true that Christ's divinity was not the subject of *Natural Religion*, Seeley was concerned there to show how Christianity "reconciled [men] to nature … by offering to them new views of the Power by which the world is governed." And in the last few pages of *Ecce Homo*, Seeley pointed out that early Christianity was essentially analogous to the youthful stage of humanity, imaginative but immature, whereas Christianity in the modern civilized world should be analogous to adulthood, having grown wise with age. In other words, Christianity had to grow and mature just as humanity itself had grown and matured. And part of that growth entailed reconciling religious belief with a form of scientific knowledge that might initially appear to conflict with it. Ultimately, this was the purpose of *Natural Religion*: to show that Christianity itself came into being as providing a new scheme of morality as well as of nature. But in the same way that its view of morality has had to adapt to the modern world, so has its view of nature. This, however, is precisely what has not happened. Christianity has largely remained stuck in a stage of youth when it comes to the knowledge of nature, while a scientific understanding of the universe has progressed independently to the point where it appears to conflict with and debunk religion. It is time for religion to come to terms with modern science. At the same time, it is important for science to recognize that, in the absence of a mature religion, it is going to have to create one.

Fundamentally, for Seeley, in the modern world religion needs to reject the supernaturalism that seemed, in its early history, to be central to its origin but is now completely out of step with modern civilization. But Seeley argued that by getting rid of supernaturalism, we are left with not simply a moralism that can exist absent a religious framework. This is because, left to its own devices, a scientific view of the world is much too harsh and not imaginative enough to help humans confront some of nature's most difficult issues, such as death. This is why, according to Seeley, so many leading scientific figures seem to be caught up with recreating versions of Christianity. He did not mention any names but presumably he was referring to Huxley's agnosticism and Spencer's Unknowable.[11] As these new forms of religion are essentially cut off from the religions of the past, they are doomed to failure. What Seeley therefore sought to show was that his envisioned "natural religion" is a product of a history, that is to say organically developed from past relations of science and religion. The key thing for Seeley was understanding this historical development and then adapting it to the modern age by shedding all of the survivals of the past while embracing those traditions that contribute to the functioning of modern civilization.

So *Natural Religion* picks up where *Ecce Homo* left off, by shedding further light on the history of religion while adding further nuance to the analogy of growth and development that appears in *Ecce Homo*'s final pages. In *Natural Religion*, Seeley argued that there are three kinds of religion that contribute to the creation of what he calls "natural religion," which is basically the new and altered form of Christianity that, Seeley argued, will perfectly suit modern life. And these three forms of religion correspond to three stages of history, which bear some similarity to Comte's stages of the theological, the metaphysical, and the positive. The first is Higher Paganism, which he argued is the childhood stage of natural religion and represents its artistic realm. Primitive Christianity is the stage of youth and represents a "phase of enthusiasm and unbounded faith both in man and the Universe." Under these terms *Ecce Homo* could be considered as a work that thoroughly describes the transition between these two stages. The third stage is that of science, "when reality is firmly faced, when the sombre greatness of the law under which we live, and at the same time the limitations it imposes on us and the patience it requires from us, are manfully confessed." It is important to understand, however, that unlike Comte's stages of history, these three stages are not completely separate from one another, but overlap and build on what came before. It is essential, according to Seeley, for the new stage "to have acquired what comes latest, but also to retain and not lose what came earlier. Humanity must constantly renew its childhood and its youth as well as advance in experience. At the same time that it observes and reasons with scientific rigour, it must learn to hope with Christian enthusiasm, and also to enjoy with Pagan freshness."[12]

Seeley argued that the present moment was that of the transition between the stages of youth and manhood, between Primitive Christianity and Science. He argued that "we" are beginning now to "acknowledge that the Universe is greater than Ourselves, and that our wills are weak compared with the law that governs it, and our purposes futile except so far as they are in agreement with that law." This is a profound transition, argued Seeley. "It is throwing off at once the melancholy and the unmeasured imaginations of youth; it is recovering, as manhood does, something of the glee of childhood and adding to that a new sense of reality. Its return to childhood is called *Renaissance*, its acquisition of the sense of reality is called Science."[13]

This new sense of reality provided by science, according to Seeley, was best represented by "That Eternal Law of the Universe," by which Seeley seemed to be referring to the law of development as applied to the cosmos, that for all things there is birth, growth, and death. This was analogous, Seeley argued, to a Judaic conception of God as an infinite power that underlies all things.[14] What was necessary, therefore, was understanding that this connection between modern science and the

Judaic conception of God existed in order that it "might form the basis of a great religion if only it revealed itself by evidence as convincing to the modern mind as that of miracles was to the mind of antiquity."[15] The way to do that, Seeley argued, was to recognize that, while science may indeed replace miracles, it actually restores the practice of prophecy, not unlike the way Christ and John the Baptist restored the practice as discussed in *Ecce Homo*.[16]

Applying the eternal law of the universe to human history, Seeley argued, could explain "the present state of affairs … and point out the direction of progress." Indeed, what was central to religious belief in the past was the construction of "[a] grand outline of God's dealings with the human race, drawn from the Bible and the church doctrine, a sort of map of history, [that] was possessed by all alike." As this old grand outline grew "unserviceable" in the present, massive "bewilderment" has ensued, as "no new map" has been furnished to take its place.[17] It was therefore absolutely necessary for Seeley's natural religion to produce a new grand outline, one furnished by the prophetic spirit found in the scientific theory of gradual development.

> As it grasps human affairs with more confidence it begins to unravel the past and with the past the future … History and prophecy belong together. As it was prophecy that made the old Church modifiable by preparing it to understand each new time, the modern Church may recover the power of development by calling history to its aid. That view of history as a whole which past generations had … may seem crude, but some such general view we must have if mankind is to be saved from bewilderment and anarchy, an anarchy which is already almost upon us. Such a view grows every year fuller and more distinct through the labours of scientific historians, a view of the past from which the future in some of its large outlines may be inferred. And thus as science replaces the old cosmologies of old religion, history scientifically treated restores the ancient gift of prophecy, and with it may restore that ancient skill by which a new doctrine was furnished to each new period and the old doctrine could be superannuated without disrespect.[18]

Seeley was, therefore, not only seeking to create a new religion. He was also seeking to provide a historical narrative that would provide the basis for this new religion, "a grand outline" that would replace the one drawn from the Bible that is no longer useful. It was, therefore, his newly adopted discipline of scientific history that would show the way, by renewing an ancient form of prophecy that would now be informed by the science of historical development. "By reviving prophecy in its modern form of a philosophy of history," Seeley argued, "we at once adapt religion to the present age and restore it to its original character."[19]

It may seem that we have come a long way from the concerns of *Ecce Homo*, but there was clearly some continuity between the two books. If *Ecce Homo* was about the moral and ethical foundations of the Christian society through a historical analysis of its founder, *Natural Religion* took the historical development of Christianity up to the present in order to show that the supernaturalism that was largely ignored in *Ecce Homo* could be safely jettisoned in the present, an entirely natural and organic development of a historical religion that was effectively merging with civilization itself. Together the two books present a historical narrative of the development of religion, one that attempts to show that that development does not end in secularization but rather a form of life that is devoted – as Goethe would have it – to beauty, goodness, and truth. Taken together, the books were an attempt to map out a grand narrative that could act as a guide in the future.

With that said, the tone of the two books *was* quite different. Maurice Cowling has put this well, arguing that in reading *Ecce Homo* one feels "that a modern mind was making itself intelligible to other modern minds without losing sight of Christ and Christianity. In *Natural Religion*, by contrast, both Christ and Christianity had flown away."[20] At the time, readers seemed genuinely surprised at just how heterodox Seeley's theology now appeared. For some, it was as if their worst fears about what *Ecce Homo* actually meant were coming to fruition, while for others it was simply evidence that the author's views had evolved, and radically so. For just about everyone, however, the book was a decidedly more solemn affair, absent the "enthusiasm" that was so central to the first volume, which was now replaced by a general dreariness about the current state of religious thought and debate. Indeed, while opinion varied about just what Seeley's message was in *Natural Religion*, there was little doubt that it marked a strong departure from the tone – if not the content – of *Ecce Homo*. If, by designating the book "by the Author of *Ecce Homo*," Seeley and Macmillan were hoping to remind readers of the experience they had when they read the original volume, such a reminder only worked to disappoint them when they read the new book and failed to be swept away by the enthusiasm for Christ's love of humanity that was so central to *Ecce Homo*.

As we've seen in previous chapters, it is fairly safe to say that *Ecce Homo* received an extremely wide but mixed overall reception. At the core of the criticism was the sense of ambiguity in the author's arguments and intentions, an ambiguity that many thought would disappear with the publishing of the sequel. Some assumed that the sequel would prove that the author was essentially an atheist (a wolf in sheep's clothing), others thought it would prove he was orthodox (a sheep in wolf's clothing), while still others assumed it would prove that he was ultimately offering a way beyond the traditional divisions within the Anglican Church. Interestingly, for many reviewers of *Natural Religion*, the ambiguity of *Ecce Homo* had apparently disappeared. Indeed, the meaning of *Ecce Homo* now seemed only too clear. But it was *Natural Religion* that was offering up confusion and ambiguity to a project that had lost its way.

Sixteen years elapsed between the times *Ecce Homo* and *Natural Religion* were published, although Seeley had formulated most of *Natural Religion* in 1875, which was just ten years after *Ecce Homo* originally appeared. The *Quarterly Review*, unsurprisingly, was terribly disappointed. "For sixteen years we have been waiting for the fulfilment of the promise held out in 'Ecce Homo,' that 'Christ, as the creator of modern theology and religion, will make the subject of another volume,' and at last we are put off with a farrago of science and culture, a pseudo-religion, from which Christ and God have been ejected to make room for Humanity and Nature. Instead of the bread we hoped for, a stone has been thrown to us; instead of a fish we have been mocked with a serpent. The inference, we fear, is inevitable, that the author's own faith has meanwhile receded."[21] Perhaps W.T. Davidson, the *Quarterly*'s anonymous reviewer, should have read the original *Quarterly* review of *Ecce Homo*, as his own interpretation of *Natural Religion* was actually anticipated by the original *Quarterly* reviewer. But this goes some way to suggest that, by 1882, *Ecce Homo* had begun to be viewed in a much less sceptical light. The original *Quarterly* reviewer believed that the author of *Ecce Homo* had no *real* faith, and that his unbelief was hidden behind a Broad Church rhetoric that could only lead readers down the same path towards outright atheism. However, in Davidson's inference that Seeley's faith had "receded" since the publishing of *Ecce Homo*, there is an assumption that he had some sort of faith in the first place. Moreover, Davidson's review goes some way to suggest just how much had changed in the intervening years, given that *Ecce Homo* was now being looked upon, by the very publication that denounced it, with somewhat nostalgic eyes. And the *Quarterly Review* was not alone in this regard.

J. Robinson Gregory, writing in the *Wesleyan-Methodist Magazine,* simply believed that the author of *Ecce Homo* had now lost his faith, arguing that "the sad, but not surprising change of position from *Ecce Homo* to *Natural Religion* manifests the insecurity of the anchorage of those who forsake the ancient faith concerning the Person and Work of Jesus Christ."[22] For Gregory, the author of *Natural Religion* "is distinctly a representative man" of the current age. "[F]rom him we can learn the direction in which the thought of an influential section of our fellow-countrymen is trending."[23] And for Gregory, *Natural Religion* represented a movement that pointed in a direction distinctly away from Christianity towards a scientific naturalism that bordered on outright atheism.

While both Gregory's and Davidson's views must be taken with a grain of salt, given that they were published in fairly conservative publications, it must be noted that even those who sympathized with *Natural Religion* tended to share the opinion that not only did the book represent a decisive shift from *Ecce Homo,* it also represented a larger shift that had affected society as a whole. For the *Athenaeum,* "it shows how fast and how far the world has been drifting since 1866 to reflect that this book takes the place of an exposition of 'Christ's theology' promised in the preface of 'Ecce Homo.'" The *Athenaeum* made this realization despite arguing that the work, much like its predecessor, was the product of a genius. At the same time, the reviewer found the tone of the book ultimately "depressing."[24] This was because the enthusiasm of *Ecce Homo* had now been replaced with a rationalism that would immensely enlighten any reader but would certainly fail to excite or renew a sceptic's faith, as was ostensibly the purpose of *Ecce Homo*. "His words are wise but sad," argued the *Athenaeum;* "it has not been given him to fire them [readers] with faith, but only to light them with reason. His readers may at least thank him for the intellectual illumination, if they cannot owe him gratitude for any added fervour."[25] Similarly, J. Llewelyn Davies believed that the "remarkable" book could "renew our hope for our country and our race," but only after "first putting us through the experience of a wholesome depression." "[W]hatever profit may be derivable from [*Natural Religion*]," Davies argued, "there is no class of readers to whom it professes to offer comfort."[26] An otherwise very positive review in the *Modern Review* ended on a similarly depressing note. *Natural Religion,* the reviewer argued, seemed richer in its suggestiveness on every read, "and more impressive in its earnestness and serious courage, more searching in its criticisms of life.

And yet the feeling of disappointment and misgiving does not pass away." That feeling of disappointment was manifested for the reviewer in a nagging doubt that the naturalism Seeley seemed to be promoting "will ever exercise the true controlling power of religion."[27] The enthusiasm that was central to the religion of *Ecce Homo* did not seem to be present anymore.

This view was echoed by G.A. Simcox in the *Nineteenth Century*, who found "nothing of the abounding buoyancy of conviction which made *Ecce Homo* rather oppressive to readers who were not carried away by it." While Simcox argued that the author generally believed in substance "all or almost all he held" when he wrote *Ecce Homo*, he argued that Seeley no longer felt as if those beliefs "were the key to everything; he has come to feel that science and nature have their rights as well as morality."[28] Seeley's new work reflected a growing desire in society that could no longer be fulfilled, namely that "we want to have everything at once" and to be shown that "each of our lives is to be an harmonious summary of the whole evolution of our planet."[29] In this way *Natural Religion* signified this desire as well as its impossibility to be fulfilled, particularly in light of the heat-death hypothesis that was gaining traction at the time.[30] For Simcox, the changing meaning of existence that was being brought about by such shifts in the physical sciences provided a larger context for understanding Seeley's narrative. And even while Seeley clearly sought to spare the reader from providing its logical conclusion, it was one the reader could not help but deduce for him- or herself, which is to say not only that there is no life after death but that this world itself may be coming to an end very soon with the burning out of the sun.[31]

Simcox alluded to the fact that the author of *Natural Religion* seemed to contradict himself in places, as if even the author did not quite believe that he had in fact harmonized the religious and the scientific, or the supernatural and the natural, to use Seeley's terminology. This problem was particularly pronounced in later reviews. "The lapse of two decades," argued W.S. Lilly in the *Dublin Review*, "does not seem to have tended to clarify or settle his [Seeley's] religious opinions." Lilly found that *Ecce Homo* was much more conclusive than *Natural Religion*, and unless Lilly forgot how ambiguous was that argument, this was saying something. "'Ecce Homo,' indeed, is plain and conclusive compared with 'Natural Religion.'"[32] The anonymous reviewer for the *London Quarterly Review* argued that "the whole style and treatment are different," in comparing *Ecce Homo* with *Natural Religion*. Sounding much like the reviewers of Seeley's *Life and Times of Stein*, the *London Quarterly Review* found *Natural Religion* "often discursive, and sometimes to all appearance aimless in its parenthetical disquisitions." While the reviewer admitted that there was a "wealth of information," he found it so "abundant that the reader willingly loses himself in such company, certain that whenever he pleases the author can unravel the tangled thread; certain, too, that presently it is sure to be tangled again."[33] Gregory said as much as well, arguing that the "aim and purpose of the book" was simply unclear, and he cited several other of *Natural Religion*'s "ablest critics" who felt the same way.[34]

One of the most critical reviews appeared in the weekly that had been central to the original reception of *Ecce Homo* as well as to the discovery of its authorship, namely the *Spectator*. Whereas Hutton originally argued in the *Spectator*'s pages that *Ecce Homo* was a remarkable study of the life of Christ despite the fundamental flaw of the author's methodology, he was unable to offer even the slightest positive commentary in regard to its sequel. He argued that "the author of *Ecce Homo*, in the hope of winning back the irreligious to religion, has refined away 'natural religion' till it comes to mean no religion at all." We can, therefore, lump Hutton and the *Spectator* together with other conservative commentators who found that Seeley had shifted so far towards the naturalist position that there was no space left for any supernatural conceptions of religion at all. "If we cannot find in the Universe as it is honestly contemplated by the mind of man, a super physical origin for physical laws, a distinctly moral origin for moral laws, and a spiritual ideal which is not of our own making, but which is imposed upon us by our Creator, we cannot find natural religion at all."[35]

Seeley did not wait, as he had done with *Ecce Homo*, to answer his critics. When Macmillan ordered another printing in August, two months after the first edition appeared, Seeley wrote a "Preface for the Second Edition" in order to clarify what he set out to do in *Natural Religion* and explain how that related to his own personal religious beliefs. What so many failed to realize, much as they had with *Ecce Homo*, was that this was ultimately an experiment to take on a particular perspective, to engage with a particular voice, that is the voice of current expounders of science. "As I meant the inquiry to be serious, I thought it essential to take the scientific view frankly at its worst. I therefore make no attempt to show that the negative conclusions so often drawn from modern scientific discoveries are not warranted, but admitting freely for argument's sake all these conclusions, I argue that the total effect of them is not to destroy theology or religion or even Christianity, but in some respects to purify all three." For this reason, Seeley admitted being surprised that his faith was once again called into question. And on this topic he was unambiguous. He stated that he had always felt that his ideas and his faith were Christian. Knowing that such a statement would not likely have been enough, he went on to announce that his ideas were, in a word, "*Biblical*," meaning "that they are drawn from the Bible at first-hand, and that what fascinates me in the Bible is not a passage here and there, not something which only a scholar or antiquarian can detect in it, but the Bible as a whole, its great plan and unity, and principally the grand poetic anticipation I find in it of modern views concerning history."[36]

Perhaps a more illuminating response appeared in the pages of the *Spectator* in the form of a letter to the editor, which was actually an excerpt of a letter Seeley had earlier written to a correspondent who was, like many of Seeley's reviewers, confused by his seeming embrace of science at the expense of the supernatural. Seeley's response to his correspondent indicates that he answered criticism of *Natural Religion* in much the same way that he did for *Ecce Homo*. He argued that there was a strategy that underpinned the way the book was written, that it was directed at a particular audience, and that therefore it took up certain concepts and arguments in order to speak directly to this audience. He was seeking to address those who embraced the scientific world view wholeheartedly and who determined to base their religious faith on scientific belief alone. He was essentially seeking "to win the attention of unbelievers." And he could only do this by accepting their faith. But the book should very much be read as a strategic experiment. Seeley wanted to find out just what would be left of religious belief "if all the negations of the fashionable scientific world were true." This was an important experiment, argued Seeley, because the forces of scientific unbelief were growing in strength such that "the men whose minds are in this state are now all-powerful over opinion, and they are forming a vast school of young crusaders, whose one ambition is to destroy religion."[37] Seeley argued, however, that just as this all-powerful group was seemingly ready to destroy religious belief, they seemed to be "seized with a misgiving. They begin to stammer out that it is not religion that they hate, but only Christianity," and that once Christianity is destroyed something will take its place.[38] "I try to catch them in this mood. I ask them to tell me what religion they will substitute. Now if it appears that this religion, is, after all, a good deal like Christianity, is not this result such as ought to be welcome to Christians?"[39]

So what did the "fashionable scientific world" think of *Natural Religion*? If their silence is any indication, it appears that they did not find it worthy of much engagement at all. This is somewhat surprising, given that Seeley perceptively exposed many similarities between naturalistic and theological conceptions of science, such as the belief of the uniformity of nature and its laws, the recognition of the inherent limitations of human knowledge, and the resemblance between what men of science refer to as nature and the Hebrew God. These are precisely the similarities that Matthew Stanley considers in comparing the theological science of James Clerk Maxwell with the scientific naturalism of Thomas Henry Huxley.[40] But these were connections that the scientific naturalists would have wanted to diminish rather than celebrate. Moreover, as many reviewers made clear, Seeley's conception of science was so broad that he made it easy for those who were already hostile to Christianity to dismiss his natural religion as underpinned by a false understanding of science itself. It is also worth noting that the figures imbued with the most cultural authority in Seeley's scheme were not the men of science but rather the scientific historians, who upon embracing a developmental view of life would be charged with predicting the future course of history.

With that said, at least one prominent Victorian intellectual found Seeley's argument for a "natural religion" entirely convincing: Frederic W.H. Myers. Myers is perhaps best known as being, along with Edmund Gurney and Seeley's old friend Henry Sidgwick, a founder of the Society for Psychical Research. Like Seeley, Myers had been engaged in seeking to bring about a rapprochement between science and religion, a rapprochement that would make religion conform with science without making it reducible to scientific naturalism.[41] For Myers and many other psychical researchers, the further progress of science into the realms of what Myers called the "supernormal" world help bring about an answer to many of the world's metaphysical mysteries.[42] It was therefore in the pursuit of psychical phenomena that science and religion could be brought together to help make sense of the nature and future of humanity.

Myers liked *Natural Religion* because it proved to "the earnest, but divergent, schools of modern thought, to the artist, the Positivist, the man of science, the orthodox Christian, that their agreement lies deeper than their differences, that the enemy of all is the same; that for the most part they are but looking at different sides of the shield, whether they worship the Unity of the Universe by the cold silver light of His power and reality, or in the golden radiance of His love." In what was an extremely positive review in the *Fortnightly Review*, Myers argued that society would do well to follow the author's lead by bringing together the truths of science, art, and religion and establish a truly "world-wide Church of civilisation."[43]

It is unclear, however, if Myers's support would have been what Seeley was hoping for.[44] To say that Myers was a heterodox Christian would be an understatement. And while his Society for Psychical Research attracted respected intellectuals such as Sidgwick and Alfred Russel Wallace, they and many of their colleagues in the society had long ago abandoned established Christianity.[45] If we take Seeley at his word, he wanted to convince both orthodox Christians and extreme naturalists that their common ground was Christianity itself, not some Spencerian "unknowable." Indeed, as we have seen, one of the main criticisms of the book was that it led precisely to the kind of non-Christian metaphysics on offer by supporters like Myers, whose review of *Natural Religion* would have only provided further evidence for that criticism. Moreover, Seeley's concerns were very much with this world, and he would have found Myers's focus on "the other world" beyond to be an unhelpful gesture that would confuse the role of Christianity in the here and now. As he argued in the letter to the *Spectator*, "My opinion in general about a future life is that we ought to believe in it, and then think as little about it as possible ... I am so full of the bearings of religion on life, society, and politics, that I find it hard to do justice to what treats of death, not life."[46]

Unfortunately for Seeley, readers did not find that enthusiasm for life, for humanity, that had been so infectious in *Ecce Homo*. *Natural Religion*, therefore, ultimately brought about a rather sad conclusion to the *Ecce Homo* controversy. *Natural Religion* was certainly not ignored as it had been when it appeared as a series of articles in *Macmillan's Magazine*, but Seeley was surely disappointed with the mixed reception of his scheme for bringing about a reconciliation between science and religion. Macmillan was more positive, believing that "on the whole *N.R.* is having a fair reception." That said, he admitted that he was somewhat surprised that certain readers did not seem to understand or sympathize with Seeley's argument. And for this he offered little explanation: "That people one felt sure would understand and sympathise, do not sympathise or understand is only a disappointment such as one must count on in this perplexed world."[47]

In terms of sales, the book initially had a modest run of success. In the first year it sold steadily, reaching four thousand copies by the end of

December 1882.[48] It was, moreover, published simultaneously by Roberts Brothers in Boston.[49] But the sales slowed considerably after the issuing of the fourth thousand in the UK. Between 1884 and 1890, the book sold on average just fifty to sixty copies per year.[50] Discussion of any further printings did not take place until 1890, at which point *Natural Religion* looked much less successful than it did in 1882. And against the sales of *Ecce Homo,* which by then had reached twenty editions, *Natural Religion* looked like an outright failure. "I suppose the sale of it [*Natural Religion*] is very much less than that of Ecce Homo," Seeley wrote to Frederick Macmillan in 1890. This was clearly disappointing. Seeley was willing to lower the price to six shillings in order to help improve the sales of the next printing, but he was not interested in revising the work in any way to improve the sales. As he told Macmillan, he had "no desire to make any alteration in it."[51] He was clearly willing to accept the disappointing sales to avoid engendering the same kind of sensational reception that awaited *Ecce Homo.* Unlike *Ecce Homo, Natural Religion* was published in a way to suit Seeley's desires. There was no vast literature speculating about who the author might be, no doubt because everyone knew that it was the Cambridge professor of history. Myers, for instance, referred to the "thin veil of anonymity criticism is still bound to respect," while Simcox argued that "it is decorous to respect his incognito, as he formally maintains it, transparent as it has become."[52] The reviewer for the *Athenaeum* all but did away with such respect by referring to the "unknown" author as Matthew Arnold's "Cambridge rival."[53] At the same time, even though Seeley was well known as the author, his name stayed out of the British periodical press in critical commentaries of the book.[54] Moreover, he would have also avoided embarrassing conversations or worrying about whether or not he could trust particular knowing friends, as there would no longer be a secret to keep.

Seeley also made sure that even after his death, *Natural Religion* would continue to be published according to the same spirit. This is most apparent in a letter addressed to Macmillan and Co. that was written not by Seeley but by his widow, Mary. Both Seeley and Macmillan had been dead for several years, and it was suggested to Mary by Dr Stanton Coit that she should inquire about having a sixpenny edition of *Natural Religion* published alongside the new sixpenny edition of *Ecce Homo.* Mary was Seeley's sole beneficiary and was also the agent in charge of his literary estate.[55] Even though the topic had not been broached by Macmillan and Co., she wanted to let them know in advance that *Natural Religion* was not to appear in a cheap edition. "I ought to have thanked you before for sending me copies of Johns sixpenny edition of 'Ecce Homo,'" Mary wrote to one of Macmillan's successors, "with which I was much pleased." She went on to say, however, that "'Natural Religion' is quite another matter. It was written for a certain class i.e., *educated* people who have a difficulty in accepting revealed religion, *not* for the general public."[56] It turns out that Macmillan and Co. agreed with Mary that a sixpenny edition should not be published, though they did so for an entirely different reason, which speaks to the irony of how the divergent interests of publisher and author can often converge. "[S]uch an edition," according to the press, "was unlikely to meet with an adequate success and is therefore undesirable."[57]

Natural Religion may have been conceived as a sequel to *Ecce Homo*, but that does not mean that it was an attempt to recreate the controversy that surrounded the publishing of the earlier volume. Seeley wanted more than anything to avoid a repeat of such an event, and this informed the way in which *Natural Religion* was serialized and later marketed as a book, one that would simply not reach a wide audience as had *Ecce Homo*. This likely was also behind Seeley's remarkable and sudden turn against popular histories written in the style of Macaulay and Carlyle, which found explicit expression most notably in his other sequel to *Ecce Homo*, the *Life and Times of Stein* (see chapter 11).

Chapter Fourteen

Remembering the Author of *Ecce Homo*

In early life he became famous, without ever acknowledging the authorship, as the writer of *Ecce Homo*, an earnest plea for natural religion, which would not nowadays excite the animosity it then provoked.

– "Sir John Seeley," *National Review* (February 1895)

By the end of the nineteenth century, the controversy surrounding *Ecce Homo* had truly become a distant memory – as had the book itself. Moreover, Seeley had seemingly transformed himself into a disinterested historian at the forefront of shaping the historical education of the nation. This may seem surprising, given his overblown reaction to a popularized history that he deemed was corrupting the nation's ability to appreciate unadorned narratives about what actually happened. But while Seeley seemed to promote a style of history that was well researched but dry and difficult for the general reader to appreciate – and here his *Life of and Times of Stein* was a case in point – his later historical work was never so narrowly conceived. That scholarship engaged with the public imagination in a way that made the past absolutely central to a seemingly ever-changing political present.

Indeed, a year after *Natural Religion* failed to live up to its billing as being written by the author of *Ecce Homo*, Seeley's Cambridge lectures on the history of the British Empire were published as the much-celebrated *Expansion of England* (1883). Even though it was by no means as popular as was *Ecce Homo, The Expansion of England* is what Seeley is now largely remembered for because of the way it seemed to catch the public mood at a moment when the British Empire was changing, and rapidly. What is rarely recognized, however, is the fact that this was an extension rather than a departure from his previous work on religious history. In this regard Duncan Bell is the exception in suggesting that the British Empire as envisioned by Seeley was an extension of his view of the development of the state itself, originating of course in the Christian society established by Christ himself.[1] Even though it was in *Ecce Homo* that Seeley asked the question, "What was Christ's object in founding the society which is called by his name, and how is it adapted to attain that object?" it was not until he wrote *The Expansion of England* that his answer was finally and fully articulated, Bell argues.

For the most part, however, Seeley's life as the professional historian of empire is seen quite separated from the time when he was at the centre of a religious controversy. It is in this way that *Ecce Homo* is mistakenly deemed a youthful literary aberration. It does not help matters that Seeley did not while he was alive acknowledge the book as his own. He quite literally made a name for himself as a modern historian of empire, and his views in this regard were met with such widespread approval among the governing elite that he was eventually knighted. Unlike his appointment to the Regius Chair of History, which was largely made on the basis of his writing a book that was still officially anonymous, he was recommended as KCMG by Lord Rosebery and the Prince of Wales on the basis of a book the authorship of which was never in doubt. In this case, he could certainly thank his old friend and publisher Alexander Macmillan, who convinced Seeley to reframe what were essentially lectures on historical methodology to be more particularly about the central example that runs throughout, namely the development of the British Empire.[2] But in doing so, Seeley became to posterity the popular historian of empire who taught that "history is past politics and politics present history." It is truly ironic that Seeley's identity has become inextricably linked with that famous dictum that he never actually said,[3] while we continue to play along with his secret by acting as if *Ecce Homo* was not really central to his life and work.

Interestingly, when Seeley died on 13 January 1895, journalists, colleagues, and friends no longer felt bound by the thin veil of anonymity that continued to protect him from being explicitly referred to as the author in discussions of *Ecce Homo* and *Natural Religion*. Obituaries, of course, appeared in all the major weeklies, monthlies, and even in some dailies and quarterlies. Seeley's recent work on the British Empire was typically central in these brief critical examinations of his intellectual life, but many commentators felt it necessary to remind the English public about other aspects of his life that were less well known. So his interest in Goethe was often discussed, as well as his youthful dalliance with

Romantic poetry along with his edition of Livy. But the central event in Seeley's life that commentators felt particularly liberated openly to discuss was *Ecce Homo* and the controversy the book had engendered. Not only did Seeley's death mark the thirtieth anniversary of the book's appearance, it also marked an end to the unspoken embargo that was placed on the public discussion of Seeley as the book's author. As we have seen, with the exception of a handful of gossip columns written once the secret of Seeley's authorship became well known, the unwritten rules of Victorian publishing etiquette largely kept his name out of the periodical press in critical discussions of the work. In death apparently such rules no longer applied. It certainly helped that during the summer of 1895, Macmillan and Co. reissued both *Ecce Homo* and *Natural Religion* under Seeley's name, but the majority of the obituaries and reminiscences appeared in January, several months before these reissues were announced.[4]

Given the controversial reception that met *Ecce Homo*, it is surprising just how views about the book and the controversy changed in light of the thirty years of distance that separated the commentators from the event in question. Most found Seeley's conception of Christ's life now to be rather commonplace, and were at pains to explain how his conception at the time was viewed as radical and also how it filled a very real need for a large portion of the reading public. The *National Observer*, for instance, argued that "those who come to read *Ecce Homo* after the author's death will probably be somewhat astonished at the contents of the work, which aroused more interest in that large part of the population who are concerned with such matters than any other book of its time."[5] The book was "neither profound nor very original," but its influence, which was great, "depended on the author's sympathy with the religious difficulties which at that time beset the persons whom he addressed." Even though the book suffers from some difficulties in logic and critical scholarship, the *National Observer* concluded that it more than anything represents "the popular state of opinion on religious matters at a time when the line of attack followed by the opponents of the system he considered himself as defending, was not as fully developed as it now is: and is itself an admirable example of the grace and self-restraint with which controversial theological subjects may be treated."[6] Similarly, but less eloquently, the *National Review* believed that *Ecce Homo* "would not nowadays excite the animosity it then provoked."[7]

For the *Saturday Review*, *Ecce Homo* was best understood as representing the state of religious thought in the 1860s, a time when "religious minded men who were serious thinkers, had found themselves forced by the march of science and criticism to reconsider their conception of the central figure and fact in the world's history." The old orthodox Christian cosmology, therefore, no longer made sense, and many under the influence of science and historical criticism were holding out hope that something would eventually take its place. According to the *Saturday*, it was *Ecce Homo* that "met the want that was so widely and deeply felt. It showed the way which thousands had been groping for, and supplied a demand which had actually existed for some time before. The ideas which it collected and reduced to order, and the feelings to which it gave articulate and sufficing expression, were no doubt in the air, and the time and the want found ... the right man."[8] The *Manchester Guardian* explained the popularity of *Ecce Homo* almost identically by claiming that it "dealt in a new way, at once bold and reverential, critical and sympathetic, with a subject of which everyone knew and felt, or thought he knew and felt, something."[9]

Maurice Todhunter, writing in the *Westminster Review*, found it similarly difficult to express how "such a moderate book as *Ecce Homo* could have caused such deep offence, even to ignorant pietists." This was in part a reflection of the great progress made since that time by "German critical ideas," but also because the very human Jesus presented in the book's pages now reflected the values most held dear. We want, according to Todhunter, "to regard 'the Man of Sorrows' as 'a human brother and friend,' instead of a distant thaumaturgist." And this is precisely what *Ecce Homo* taught readers to do, which is why Todhunter argued that "[o]nly those who cling obstinately to tradition can any longer condemn it [*Ecce Homo*] as being subversive or impious. It claimed to show, on historical grounds, how a great Personality had changed the face of the world, and introduced new motives for conduct, by the force of human enthusiasm."[10] Or, as the *Academy* would put it, *Ecce Homo* "represents, more clearly than elsewhere, the humanitarian change that has come over Christianity in the eyes of all enlightened laymen."[11]

Even though throughout the obituaries *Ecce Homo* is presented as producing "an historian's conception of Jesus,"[12] Seeley's stake in doing so is largely determined as coming from a deeply personal desire to know who Jesus really was, untethered to Anglican dogma. In this way, *Ecce Homo* was, according to the *Saturday Review*, "conceived and written to satisfy its author's own mind."[13] Herbert Fisher, writing for the *Fortnightly Review*, argued that *Ecce Homo* "was inspired by a moral rather than by a strictly scientific interest."[14] But, according to the *Speaker*, Seeley's respect of "all sacred conventions, however different from his own, and the fear of wounding the feelings of those who were near and dear to him was probably his motive in withholding his name from 'Ecce Homo.'"[15] *Ecce Homo* was a book, Fisher argued, which "was not professedly unorthodox, and yet it invited the orthodox for a moment to discard the associations of divinity, and to concentrate their gaze upon the spectacle of a perfect human life passed upon earth."[16] And yet Seeley was far removed from orthodox Christianity, and his writings bear the mark of a profound distrust of dogma. "Like many other Cambridge men of his generation," argued the *Speaker*, "he admired F.D. Maurice for the holiness of his character as well as for the part he played in destroying certain popular accretions which formed no part of the gospel of Christ. I once heard him call F.D. Maurice one of the greatest men of the century."[17] Such a statement would have surely been controversial in the 1860s, but by 1895, Maurice too was no longer considered a divisive figure. Indeed, once we remove *Ecce Homo* from the specifics of the religious controversy that it engendered, according to *The Times*, it "sheds this clear light upon the personality of the author, that his creed was one in which religious theory and daily practice formed indivisible parts of the sum of life."

The fact that the publishing of *Ecce Homo* led to a sensational controversy was, of course, often alluded to. *The Times*, for instance, argued that *Ecce Homo* "created more sensation in the cultivated world of England than perhaps any book of its kind before or since." That obituary actually spent considerable space describing the role anonymity played in the sensation, recounting how the authorship "was attributed in turn to persons differing as widely as the Archbishop of York and the Emperor Napoleon III., the Poet Laureate and George Eliot … Roman Catholic divines and Nonconformist ministers shared with innumerable laymen the undeserved honour of having written it."[18] For the most part, however, the obituaries said little about the anonymity issue, choosing instead when describing the controversy to focus on the nature of the opposition to the book. The *New York Times*, for instance, suggested that "Prof. Seeley's chief work," *Ecce Homo*, "created great excitement among the members of the various Protestant communities, and elicited numerous replies."[19] That the *New York Times* would not expand on the nature of such replies is not unexpected, but even the British press failed to add much nuance to the discussion, claiming that the opposition to *Ecce Homo* came from the evangelical set of the Church of England. The *Saturday Review* used the opportunity of Seeley's obituary to laugh once again at "Lord Shaftesbury and the Evangelical Party" who "lost no time and spared no violence of language in denouncing it [*Ecce Homo*] as unorthodox and untrue."[20] Joseph Jacobs, writing for the *Athenaeum*, argued that in laying "stress on Jesus's personal influence as a man upon men," Seeley "thereby raised the ire of the Evangelicals."[21] The *Spectator* as well, whose obituary is discussed in greater detail below, argued that *Ecce Homo* brought about an "enthusiasm of annoyance in the Evangelical party," as if all opposition to the book came from one particular Protestant sect.[22]

More unexpected, perhaps, is the treatment of *Ecce Homo*'s sequel, *Natural Religion*. Todhunter, for instance, explained that *Ecce Homo* left unanswered many important questions in regard to the life of Christ, "and many looked forward to the promised sequel." But the sequel took sixteen years to appear and seemed to deal with an entirely different set of problems. Todhunter admitted that many disparaged the book when it was finally published, but this was largely because "it appeals less to ordinary minds." As far as Todhunter was concerned, "it is in no way inferior to its more exoteric forerunner."[23] The *Manchester Guardian* also pointed out that *Natural Religion* was much less popular than *Ecce Homo* but argued that this was because *Ecce Homo* had largely been forgotten by the time it appeared. Moreover, "[t]he later book has all the excellence of style which characterised the earlier, but the subject is abstract instead of personal, and the conceptions with which it deals are beyond the reach of many minds."[24] Jacobs argued that in *Natural Religion* Seeley very much set out to extend the argument offered in *Ecce Homo* by bringing about a reconciliation between science and religion. Jacobs was not entirely convinced that Seeley alone was responsible, however, believing that the current stalemate between science and religion was likely "produced rather by a process of exhaustion than by any direct influence of Seeley's." And yet Jacobs was forced to admit that "'Natural Religion' was fully as original as 'Ecce Homo,' and was much more attractive in style."[25]

While the aforementioned obituaries and death notices must be read with some caution, given that they tend towards the hagiographical rather than the critical, one particular obituary stands out less because it was critical than because of its flippant tone. In just a few sentences, the *Spectator* obituary managed to explain the controversy of *Ecce Homo* along with its relationship to *Natural Religion* and Seeley's style of writing in a way that made the current focus on Seeley's life seem rather silly. "Sir John Seeley first attracted attention by a book," the obituary began, "'Ecce Homo,' which, though now but little read, excited thirty years ago a sort of enthusiasm of annoyance in the Evangelical party. Though not orthodox, it was by no means so un-Christian as is generally asserted, being rather the work of a man who wished to believe and expected that he should, – an expectation, however, which, as a subsequent book on 'Natural Religion' appeared to show, was hardly realised." The review went on to criticize Seeley's style of writing, which it argued was "almost unendurably tedious," whether in *The Life and Times of Stein, Ecce Homo,* or *Natural Religion.* About *The Life and Times of Stein* in particular, the *Spectator* argued that readers of the book would likely have lost any interest they might have had for the subject beforehand.[26]

As we have seen, Seeley had had a few disputes with the *Spectator* over the thirty years since *Ecce Homo* was first published. And it was the *Spectator*, of course, that finally unmasked Seeley as the book's author. It is somewhat fitting then that the *Spectator*'s brief obituary brought about a final controversy concerning Seeley's life. Many quite rightly found the review contemptuous. The *National Review* was perhaps being kind in referring to it as "an unappreciative little note."[27] And at least two attempts were made to answer the obituary in the form of letters to the editor. Only one was actually published, however, written by Alfred Church. He found the *Spectator*'s "estimate of Sir John Seeley's religious position" entirely inadequate, arguing that *Ecce Homo* was written "in the hope of arresting the decay of faith which he [Seeley] saw about him. It was teaching, he thought, of the kind that the age could bear." He went on to quote from Seeley's "An Easter Hymn," which was sung at his funeral and printed on the funeral card (see figure 14.1), to indicate just how "his own thoughts moved in an altogether higher sphere" than anything that would have appeared in *Ecce Homo* and *Natural Religion*.[28]

A much lengthier letter to the editor of the *Spectator* would not appear in the weekly's pages. It was, according to Hutton, the *Spectator*'s longtime editor, far too lengthy.[29] Its author, however, felt moved to send the letter on to Mary Seeley, hoping that she might appreciate the attempted remonstration with the *Spectator* "against their absurd notion of your husband."[30] The writer was none other than J. Llewelyn Davies, the very man responsible for confirming Seeley's identity as the author of *Ecce Homo* to the editor of the *Spectator* thirty years earlier. Davies was quite convinced that he spoke for many of the *Spectator*'s readers when he said that its obituary of Seeley was "surprisingly inadequate."

IN LOVING MEMORY OF

John Robert Seeley,

K.C.M.G., LITT.D.

Regius Professor of Modern History in the University of Cambridge, Fellow of Gonville and Caius College, and Honorary Fellow of Christ's College.

who was born September 10, 1834; and died January 13, 1895, at 7 St. Peter's Terrace, Cambridge, and whose body was laid to rest in Mill Road Cemetery, January 17, 1895.

The first three verses of the following hymn were sung at the Funeral Service in Caius College Chapel.

What avails that winter die
If death die not, winter's sting?
Hopeless, loveless, man would lie,
Crown'd not Eastertide his spring.

Who would hope at all, or strive,
Overwhelmed by fatal force?
Who would love at all, to grieve
Parted by that dire divorce?

But from yonder gulf of gloom
Not the Lord alone is risen;
Hope with Him has left the tomb,
Love with Him has burst the prison.

Saviour, for this other spring,
Scents of Eden breathing near,
Gratefully these gifts we bring,
Flowerets of the freshening year.

An Easter Hymn, J. R. Seeley.

Figure 14.1 Seeley's funeral card, Seeley Papers, MS 903A/3/3. Courtesy of the Senate House Library, University of London.

He certainly agreed that *The Life and Times of Stein* "was not an attractive piece of literature," but he suggested that it was ridiculous to say that readers interested in Stein beforehand would have lost interest by reading Seeley's rich and well-documented history. With that said, Davies was mainly concerned with correcting the *Spectator*'s treatment of *Ecce Homo*, which was, unlike *The Life and Times of Stein*, "an extraordinarily attractive book." Moreover, Davies continued, the fact that it drew derision from some members of the evangelical party was only "one small part of its effect." And here Davies could have been referring to many of the obituaries that oversimplified the response to *Ecce Homo*. It was certainly true, Davies argued, that "[i]t was felt by old-fashioned Evangelicals to be dangerous and Lord Shaftesbury expressed himself absurdly about it at Exeter Hall, – to the 'annoyance' of many of his friends." But the reason certain evangelical leaders got so worked up over the book was because many "members of the party were so much interested by it. It was a book that everybody read, and it left in almost all minds a sincere desire for more."

Davies was willing to admit that the *Spectator*'s claim, along with many other obituaries, that *Ecce Homo* was no longer read might indeed be true. And this was a shame, because "anyone in the present day who is set thinking and wondering by the Gospels would find it a profitable book to read. The style of it to my mind and ear is of quite rare excellence." It was simply unreasonable for the *Spectator* to suggest that Seeley "had no sense of proportion; and his style, when he had room enough, is, in spite of its plainness, almost unendurably tedious." According to Davies, in respect of its English, "I have been accustomed to regard Ecce Homo as … one of the choice books of its literary period. The style is plain, as you say; it is unaffected and unadorned. It is also accurate, charged with feeling, vigorous; and every sentence tells."[31] Davies's letter was a wonderful defence of both Seeley and *Ecce Homo*, but it *was* far too long as a "letter to the editor." Aside from what I have summarized above, Davies also included long excerpts from *Ecce Homo* in order to illustrate properly the beauty of the book, in contrast to the *Spectator*'s dismissive view. Lady Seeley did clearly appreciate the letter, however, as she sent it to family members in order to share Davies's commentary about her husband. But it truly was a "pity," Richmond Seeley wrote to Mary, that Davies "made his letter to the Spectator so long."[32]

As it would turn out, Mary was already at work producing a final monument to her husband's life. While Seeley would never receive the "life and letters" treatment that was typical for well-known Victorian men of letters,[33] she did see to it that Seeley's last history book would contain a relatively substantial biographical sketch of his life. Before he died, Seeley had been working on a manuscript that would be published posthumously as the two-volume *Growth of British Policy* (1895). G.W. Prothero kindly took on the role of editor and exchanged a series of letters with Mary concerning the new volume. Prothero was, by all accounts, happy to take on this role, which largely involved checking the proof sheets and arranging for the final chapter to be written from some of Seeley's lectures on the topic. "I need hardly say that I am very sensible of the honour you do me in asking me to see my late master's book through the press," Prothero wrote to Mary, "and that anything I can do that would be of service to you and be helpful toward making known his work to the world, I shall do with the keenest pleasure, though with a heavy sense of responsibility."[34] Even when editing the book began to be a bit more onerous than initially advertised, Prothero was adamant that he felt that it was "the greatest honour to have to do this work at all, and only wish I were more worthy to be his literary executor."[35]

Prothero had originally planned to contribute an article about Seeley's life for the *National Review*, and he requested that if Mary could "supply me (if you do not mind doing so) with any works of his views about education and other topics, his early life, method of work and habits that would be generally interesting, I should be extremely grateful for them."[36] Mary would end up providing much more material than Prothero expected, and at some point it was decided that Prothero's *National Review* article about Seeley would be better suited as a "Memoir" that would appear at the beginning of *The Growth of British Policy*. This "Memoir" would be the first real attempt to write a biographical account of Seeley's life and work. And some of the material Mary provided was clearly meant to counter some of what was being said about Seeley in the periodical press. In an ongoing dialogue with Mary about his brother's estate and the tricky business of royalties and future editions of Seeley's work, Richmond mentioned that "the newspapers have spoken of *Ecce Homo* as not read now." He suggested that it would be easy enough to find out if this was actually true. "I suppose you can tell about [what] the annual sale has been of late, and if it has been at all considerable I think you [should] send this figure to Professor Prothero for his article."[37]

Whether Mary acted on this advice or not is unclear, but she certainly provided Prothero with some telling facts; Prothero was particularly thankful "for the very interesting information you give me about 'Ecce Homo.' I am very glad to have it. It is a book that meant a great deal to *me* personally."[38] So it was that, as in the recently published obituaries, *Ecce Homo* occupied a central place in Prothero's "Memoir" and is described as "the best known and in some respects the most remarkable of his works." Prothero praised the book's crisp "style and limpidity of expression" as well as the novelty with which Seeley treated his subject. All this was fairly typical of Seeley's obituaries, but where Prothero differed was in his presentation of the "storm of controversy" that the book had aroused. *Ecce Homo* was, according to Prothero, written deliberately to avoid controversy, and yet it engendered controversy anyway. This was because "Its restriction of the view of Christ to the human side of his life and teaching was attacked by many as implying the non-existence of any other side. Avoidance was regarded, without warrant, as negation."[39]

Even though Prothero argues at the end of his "Memoir" that it is not a place to estimate Seeley's position as a historian, he does offer a few general remarks. In this final section of the memoir there is much said about Seeley's views of the state, about his role in treating history like a science, about making history and political science seemingly similar subjects, and, perhaps most significantly, about denying the rationale for a "little" England. Seeley's religious views and his role in shaping the religious thought of the past thirty years are not deemed terribly important, at least when it came to Seeley as a historian. Here we begin to see the beginnings of Seeley being understood only as a scientific historian, with the controversial religious studies disconnected from his later more serious labours in helping to educate the nation.

Moreover, even though Prothero has something to say about the controversy of *Ecce Homo*, the fact of the book's anonymous status is mentioned but is determined to be barely relevant, as in just about all of the obituaries written in the wake of Seeley's death. "The book was published anonymously," Prothero says at the end of his discussion of *Ecce Homo*, "but the authorship soon became an open secret."[40] This is the line that is run by every commentator since, as if no further explanations need to be offered. Anonymity was often mentioned, but unremarkably, as if that was an interesting fact about the book that actually played no role in its reception. This is the beginning of constructing the larger narrative of Seeley's life, and this trope of irrelevant anonymity is part of it.

But saying that *Ecce Homo* was published anonymously only to become an open secret is not an adequate explanation but rather a point of departure that necessarily leads to a series of questions that have never been adequately dealt with: Why was it published anonymously? How was it published anonymously? What role did its anonymous status have to do with the controversy over the book? How did Seeley's secret get revealed? If it was such an "open secret," why was the thin veil of anonymity kept for so many years? These are just some of the key questions that this book has asked, and I am aware that some of them do not have simple or clear answers, particularly the last one. But they are questions that take us into the heart of one of the most controversial debates of the century, which was not just about the historical Jesus but about the nature of authorship and authority, of anonymity and identity, of publishing and marketing.

Epilogue

Anonymous Publishing and Universal History

In these days, when it is sometimes hinted that the part played by the publisher in the production of literature is merely that of a superfluous middleman, it is well to consider the career of such a one as Alexander Macmillan.

– J.S.C., "Alexander Macmillan," *Academy*, 1 February 1896

Is there a chance of the chair being made one of Universal History?

– Seeley to Kingsley, 1869, Kingsley Papers

It is worth emphasizing, at the end of what may on the surface appear to be a fairly narrow study, that the appearances of *Ecce Homo* and *Natural Religion* mark key moments in the history of anonymous publishing. In the first half of the nineteenth century, anonymous and pseudonymous works dominated the publishing world. Seeley's father, for instance, was doing something quite common by not including an authorial signature and then later linking a series of his books under the designation of "by the Author of *Essays on the Church*." By the time *Ecce Homo* appeared in 1865, however, there were growing debates about the practice of anonymity and pseudonymity. Why would an author not stand behind his or her views? What, indeed, was such an author hiding by not signing his or her name? Moreover, as the reception of *Ecce Homo* became deeply entangled with assumptions about the author's identity, questions about the ethics of producing and marketing such an anonymous sensation became more central in critical discussions of this book more generally. The subject matter certainly played a role in intensifying this discussion, because readers had to place a certain amount of trust in the unknown author in order to be taken on a seemingly perilous religious journey. It did not help matters that the author himself claimed that he did not know where that journey would lead. And had readers known at the time that the journey would take them to the conclusions offered in *Natural Religion*, no doubt many would have declined reading *Ecce Homo* in the first place. "By the Author of *Ecce Homo*" may have been a designation meant to remind readers of the experience they had upon originally reading *Ecce Homo*, but for many that reminder was likely bittersweet, as *Natural Religion* failed to meet the expectations set by that original sensational study of Christ.

While authorial identity was perhaps nowhere more debated than in the world of theological publishing, the debate about anonymous journalism also began at roughly the same time that *Ecce Homo* appeared. And within this genre of publishing, Seeley's friends and colleagues, from Macmillan and Maurice to Hughes and Davies, were convinced that anonymity represented a bankrupt form of journalism that contributed to the growing sectarianism in British intellectual life. They argued that journals needed to produce ideas and debates that could be attached to the individuals articulating them, not to corporate identities or party ideologies. By 1875, when *Natural Religion* began to appear in serialized form in *Macmillan's Magazine*, the debate about anonymous journalism had clearly been won by those on the side of the individual signature, and yet Seeley was adamant that his articles be published anonymously, thereby contradicting the authorial trends in journalism at the time and even the editorial policy of the magazine within which they were due to appear.

While this present study set out to show that Seeley's rationale for wanting *Ecce Homo* and *Natural Religion* published anonymously was a complicated one that often relied on the same sorts of arguments that critics of the practice relied upon, namely that by removing his name he believed that he was making it possible for his interpretation to transcend particular party or corporate perspectives, by the time *Natural Religion* was published the debate was largely over and anonymous publishing in general was in sharp decline. Moreover, whereas anonymity in theological publishing was, particularly before the nineteenth century, quite common, by the end of the nineteenth century it was barely tolerated. But by then the kind of protection anonymity offered for those writing in that highly controversial genre, such as protection against religious persecution, was no longer as necessary as earlier in the century when charges of heresy and blasphemy were more common. Moreover, the main parties of the Anglican Church that were seemingly so central in shaping religious debate in the mid-nineteenth century were no longer in control of that debate by the end of the century. Theological works lacking a signature had, therefore, lost much of the social and legal context that made them necessary earlier in the century.

It follows that, as the individual author's name became more central to publishing as the century wore on, the role of publishers became less apparent in the marketplace of ideas. In the context of this study, Macmillan's marketing strategy was much more apparent during the first year of *Ecce Homo*'s publication when he was the only public face for the book than was the case when he published, for instance, *The Expansion of England* in 1883, a book that included Seeley's signature. And yet we know, thanks to Leslie Howsam, that Macmillan was a powerful influence on the early shaping of *The Expansion of England*, such that its great success must also be attributed to Macmillan's publishing and editorial expertise. But that is a story that is much harder to tell and one that at the time was completely hidden from the public view. By the end of the nineteenth century, publishers had receded from the public stage, as had their perceived influence.

We need not look any further than Macmillan's death in 1896, almost one year to the day after Seeley's death, for further evidence of this. Despite the fact that Macmillan published many of the most important works to appear in the nineteenth century, brought together many of Victorian Britain's leading intellectual figures in his "tobacco parliaments," and more than anything contributed immensely to the increase of knowledge and literature over several generations, his death received little notice in the periodical press. In contrast to the wealth of information that appears in any search of the Victorian periodical databases about Seeley's death in 1895 (as discussed in chapter 14), a similar search for Macmillan in 1896 produces, for the most part, only a few short notices, along with a "reminiscence" in *Macmillan's Magazine* that was written by a close friend.[1] There is a two-length column obituary in the *Academy*, however, which sought to explain just how important Macmillan was to a half-century of publishing. Perhaps most interesting about the obituary is the author's awareness about how the public's perception of publishers had changed by the end of Macmillan's life. "In these days," argued the author, "when it is sometimes hinted that the part played by the publisher in the production of literature is merely that of a superfluous middleman, it is well to consider the career of such a one as Alexander Macmillan. In him were combined – to an extent perhaps not equalled by any one since the first

John Murray – those qualities which sweeten and dignify business, without depriving it of the pecuniary rewards which it deserves."[2] It is worth noticing that this was not a message that was widely disseminated upon Macmillan's death.

This study has shown that, in the case of *Ecce Homo*, Macmillan was no mere middleman. Much of the book's success must be attributed to the efforts of the publisher whose marketing strategies were paramount in creating a sensation surrounding the book, even while some of those strategies in hindsight seem questionable. Moreover, despite the attention *Ecce Homo* received, it was still likely a minor success for Macmillan when considered in the context of his long publishing career. Indeed, even though Seeley was angry with Macmillan and the way his secret was eventually exposed, he clearly recognized that Macmillan had gone above and beyond what was expected of a publisher. He apparently told Macmillan that "No other publisher would have done for the book what you have done."[3] And that is no doubt true.

Much more work needs to be done to map out a larger history of anonymous publishing in the nineteenth century, one that connects the debates engendered within particular genres to what appear to be declining trends in its usage and appeal. What role, for instance, does the rise of the celebrated individual author play in bringing about a decline in anonymity? Moreover, is there any evidence to suggest that a decline in anonymous publishing relates at all to changing public perceptions of publishers and the role they play in the publishing process? While this study has not sought to answer those more general questions, I hope that the example presented of the publishing and reception of *Ecce Homo* and its sequel *Natural Religion* could fruitfully be applied to other cases in order to shed further light on the more general trends of authorship and publishing in the nineteenth century.

Finally, while this book has been about the controversy engendered by the anonymous publishing of *Ecce Homo*, it has also been about the afterlife of the book and how the memory of the controversy faded to the point that, by the time of Seeley's death, readers needed to be reminded about its very existence. It was clearly a radical book when it was published, as it dealt with the problem of the historical Jesus in a way that was entirely foreign to many of its readers. This was most apparent in the reviews, which struggled to understand the book's methodology and message while trying to imagine the identity that lay just behind the mask of anonymity. By the time *Natural Religion* was published, readers seemed to understand *Ecce Homo* only too well, but it was now *Natural Religion* that was befuddling. An attempted reconciliation between science and religion was highly desirable, but many Christian readers found that *Natural Religion* either gave too much ground to naturalism or else defined religion in such a way as to make the term almost meaningless. And yet, as we saw in chapter 14, by the time Seeley died both books seemed less radical than they had when they were first published. The passage of time surely softened the blows. And now that they have largely been forgotten, it is difficult to imagine just how central they were to the rapidly transforming religious debates of the second half of the nineteenth century.

But they clearly were central. Indeed, if *Ecce Homo* can be understood as representing a moment when a humanistic understanding of Jesus Christ was established to help bring together a wide range of theological perspectives, *Natural Religion* represents a very different moment, when the kind of appeals made in *Ecce Homo* had come to feel quaint and rather orthodox. It is clear that by the time *Natural Religion* appeared, Victorians had largely internalized the character of Christ as presented in *Ecce Homo*. But the conclusions that Seeley made in the earlier work about how his humanistic conception of Christ could modernize Christianity were no longer convincing. For instance, the debate Seeley had with himself in *Ecce Homo* about whether miracles were actually wrought would not have been taken seriously in *Natural Religion*, as Seeley found it necessary for religion to abandon supernaturalism entirely. At the same time, he grasped only too well that some naturalist critics of Christianity were recognizing that religion of some kind was necessary in a world without Christianity, and the publishing of *Natural Religion* represents a moment when religions of the future were being imagined to occupy that space.

It has become a truism that the nineteenth century was an age of transition. And, intellectually speaking, nowhere were contemporaries more aware of this transition than in the realm of religious thought. By considering the receptions of *Ecce Homo* and *Natural Religion*, we gain much insight into the nature of that transition, the pace of which clearly quickens during the sixteen years separating the two works. It is, of course, possible to argue that the publication and reception of the two works simply provide further evidence for the so-called secularization thesis, which argues that as scientific advancements challenged traditional religious doctrines, religious belief receded from the centre of life to the background, thereby allowing for secular ideas and institutions to merge effectively with modern civilization. And, indeed, one of the ways to think about the transition from *Ecce Homo* to *Natural Religion* is precisely one of secularization: Seeley began with a study of Christ's character but concluded with a plea to reject the supernatural foundation of Christianity in order to preserve some form of religious belief in the modernizing world.

Seeley, however, wrote each work with a very different design in mind in terms of what it says about the state of contemporary religious belief. While his *Natural Religion* was often referred to as an attempt at creating a "secular" religion, Seeley wanted more than anything to combat the growing threat of secularization that he witnessed towards the end of the nineteenth century. Recall that he argued at the end of *Natural Religion* that what was now missing from the modern world was a grand cosmic narrative like that produced by the Bible that could give shape and meaning to a society in rapid transition. That particular biblical narrative was no longer convincing to the scientific minds Seeley hoped to reach, but he did not advocate the construction of a new secular myth. He argued that his narrative was decidedly religious – particularly Christian – and adapted to the realities of modern civilization. *Ecce Homo* and *Natural Religion* can therefore be read together as an attempt to construct part of that new religious narrative, one that began with a description of the emergence of Jesus Christ and the construction of a moral framework that was, when properly understood, completely adaptable to the humanistic and scientific ideals of the nineteenth century. It ended with an understanding that the religion produced by this moral framework was a naturalism that only became apparent with the rise of modern science and the decline of the unnecessary ancient survival that was supernaturalism. Progress itself was understood as being a product of the further merging of the religious and the natural, while secularization was equated with social decline.

Seeley was therefore seeking to construct a universal history, one that relied on scientific history to resurrect the practice of prophecy while also still embracing the providential framework that was so central to universal histories of the past.[4] Seeley made this clear at the end of *Natural Religion* when he said that he wanted to revive prophecy in the modern form of a philosophy of history, thereby "adapt[ing] religion to the present age and restor[ing] it to its original character."[5] This is important to note because it contradicts one of the main historiographical tropes of nineteenth-century historical writing, namely the claim that "professional historians expelled universal history from the discipline" in favour of narrow nationalistic histories of modern states.[6] Seeley's

Expansion of England and *The Life and Times of Stein* are often invoked as particularly powerful examples of this move. But when these works are placed within the context of the grand narrative of Christianity and politics that appears in *Ecce Homo* and *Natural Religion*, it becomes clear that his seemingly nationalist histories were also parts of this larger project of universal history.

In this regard, it is suggestive to recall (see chapter 11) that when Kingsley approached Seeley about the possibility of replacing him as Regius Professor of Modern History at Cambridge, Seeley felt himself incapable of taking the position. In part this was because he was largely trained as a classicist and had not, therefore, examined any of the original authorities of modern history. He would have no problem taking the position if the title could be altered, however. "Is there a chance of the chair being made one of Universal History?" Seeley asked Kingsley. "Gladstone, I suppose, can do what he likes with it. If ancient history were included I should not feel so diffident."[7] We know, however, that Seeley would go on to accept the position and produce works of modern history that relied extensively on just those original authorities he had yet to examine. But it must be understood that Seeley's "modern" studies were never so narrowly conceived as is typically understood in studies of nineteenth-century national histories. His were always situated within an implied universal history that began in an immense age of transition with the arrival of Jesus Christ and ended in another immense transition when Christ's society and enthusiasm for humanity would have to be adapted to a new social and political reality. *Natural Religion* was thus an attempt to show how Christianity could be adapted to a modern naturalism. Seeley's studies of the British Empire and Germany, therefore, must be understood as attempts to illustrate that the history of the modern state was yet another part of the story that he sought to tell about the founding and development of Christ's society.[8]

This study, therefore, undermines the view of Seeley as being representative of a modern, secular historian of the new nation-state. While he did argue that history needed above all to be concerned with the history of politics, his view of politics was never just about the rise of modern secular nation-states. The driving force behind Seeley's historical work was to show how the development of religion and the state were parts of the same process of civilization and that future progress was therefore dependent on ensuring their further integration. The implication was that it was scientific historians such as Seeley who would be the new modern prophets, whose historical, religious, and political knowledge could help indicate the path of future progress. This is quite in contrast to the prominent view of the nineteenth-century political historian as a specialized, dry-as-dust fact grubber who has closed himself off from the real world by taking shelter in the ivory tower of the academy. Seeley never quite fitted that model. He was first and foremost a practitioner of universal history. And *Ecce Homo* was the foundation for that universal history.

Notes

Prologue: The Forgotten Story of Ecce Homo

1 This story is recounted in [A.P. Stanley] APS, "Ecce Homo," *Macmillan's Magazine* 14:8 (June 1866): 134; and Morgan, *The House of Macmillan (1843–1943)*, 79.
2 Matthew Arnold to Mary Penrose Arnold, 23 February 1866, in *The Letters of Matthew Arnold*, vol. 3: *1866–1870*, ed. Cecil Y. Lang (Charlottesville and London: University Press of Virginia, 1998), 14.
3 "A Popular Religious Work," *Bury and Norwich Post and Suffolk Herald*, 17 April 1866, 3.
4 See the advertisement for the "Nineteen Thousand" printing of *Ecce Homo* in the *Spectator*, 11 January 1868, 59.
5 Howsam, *Cheap Bibles*.
6 Christian Fraser, quoted in "Spanish Fresco Restoration Botched by Amateur," accessed 16 April 2015.
7 See the "Reviews of *Ecce Homo*" section of the Bibliography.
8 J.R. Seeley, *The Expansion of England: Two Courses of Lectures* (London: Macmillan, 1883).
9 See, for instance, Koditschek, *Liberalism, Imperialism, and the Historical Imagination*; Bell, *The Idea of Greater Britain*, ch. 6; and Hesketh, *The Science of History in Victorian Britain*.
10 On Seeley's "political theology of empire," see Duncan Bell's *Remaking the World*, ch. 11.
11 Pals, "The Reception of 'Ecce Homo'" and *The Victorian "Lives" of Jesus*, 39–50, 57–8. However, buried in a footnote (ibid., 57n96), Pals comments on the fact that Dean Church believed the author to be John Henry Newman. He reasons there, but not in the main text, that perhaps this had something

to do with Church's surprisingly sympathetic review of *Ecce Homo* in the *Guardian.* This review receives extensive treatment below (see ch. 5).

12 Wormell, *Sir John Seeley and the Uses of History*, 21–38.

13 This particular trope seems to have originated in the periodical literature at the time of Seeley's death (discussed in ch. 14), which referred to Seeley's authorship of *Ecce Homo* as an "open secret." It was picked up by the secondary literature on Seeley soon afterwards. See, for instance, Benn, *The History of English Rationalism in the Nineteenth Century*, 2:236.

14 The Papers of John Robert Seeley, Senate Library, University of London, London (hereafter Seeley Papers), MS 903/3A/1.

15 For an excellent introduction to the interdisciplinary approach to book history, see Howsam, *Old Books and New Histories*.

16 Secord's book is just one among many important recent histories of science that have been written from the perspective of the history of authorship, reading, and publishing. See especially Johns, *The Nature of the Book;* Topham, "Beyond the 'Common Context'"; Fyfe, *Science and Salvation*; and Fyfe and Lightman, eds., *Science in the Marketplace*.

17 Lightman, *Victorian Popularizers of Science*.

18 Secord, *Victorian Sensation*, 515.

19 Another similar study that unfortunately appeared in the late stages of writing this book is Priest, *The Gospel According to Renan*. Priest also examines the reception of a single book, Renan's *Vie de Jésus* (1863), and does a particularly remarkable job exploring the way the public interpreted the book by examining the hundreds of letters Renan received from general readers. While Seeley also received many letters from general readers, only a few of these have been preserved in Seeley's archive. Much of the evidence utilized in this study to get at the reception of *Ecce Homo,* therefore, is limited to the extensive review literature.

20 This is the argument put forward by Adams and Barker, "A New Model for the Study of the Book." Adams and Barker were criticizing the famous model of Robert Darnton's classic "What Is the History of Books," which argued that the processes of producing a book begin with the author. For a useful summary of this debate, see Howsam, *Old Books and New Histories*, 28–38.

21 Howsam, *Cheap Bibles;* Howsam, *Kegan Paul, a Victorian Imprint*; and Howsam, *Past into Print*. See also Aileen Fyfe's recent *Steam-Powered Knowledge*.

22 On Macmillan's shaping of historical writing, see Howsam, *Past into Print*, 24–40, 43–62; and Howsam, "Academic Discipline or Literary Genre?" On Freeman, see Bremner and Conlin, eds., *Making History*.

23 That said, Robert L. Patten's John Coffin Lecture of 2006, "Anon.," explored some of the commercial reasons for publishing anonymous fiction, such as the desire to create a buzz about a given book. I am grateful to Professor Patten for sending me a copy of this wonderful lecture.
24 See, for instance, Griffin, "Anonymity and Authorship"; Griffin, ed., *The Faces of Anonymity*; Mullan, *Anonymity*; Ferris, *The Achievement of Literary Authority*; and Patten, *Charles Dickens and "Boz."*
25 See Patten, "Anon."
26 Hadley, *Living Liberalism*, ch. 3. See also Gowan Dawson, Richard Noakes, and Jonathan R. Topham, "Introduction," in Cantor et al., *Science in the Nineteenth-Century Periodical*, 1–36, on 20–1.
27 J. Llewelyn Davies to Alexander Macmillan, 13 November 1866, Seeley Papers, MS 903/3A/2.
28 Buurma, "Anonymity, Corporate Authority, and the Archive."
29 Wormell, *Sir John Seeley and the Uses of History*, 15.
30 See, for instance, Morgan, *The House of Macmillan (1843–1943)*.

1 Authority and Authorship

1 Henry Sidgwick to his mother, 29 January [1866], in Arthur Sidgwick and Eleanor Mildred Sidgwick, eds., *Henry Sidgwick: A Memoir* (London: Macmillan, 1906), 140.
2 Sidgwick to his mother, 19 February 1866, ibid., 143.
3 *Ecce Homo: A Survey in the Life and Work of Jesus Christ*, 5th edn (London: Macmillan, 1866), xxi.
4 The secondary literature on this issue is extensive. But see, in particular, Moore, *The Post-Darwinian Controversies*; Bowler, *The Non-Darwinian Revolution*; and Lightman, "Darwin and the Popularization of Evolution."
5 Burrow, "The Uses of Philology in Victorian Britain," 180.
6 On the *Essays and Reviews* controversy, see Ellis, *Seven against Christ*; Altholz, *Anatomy of a Controversy*; and Shea and Whitla, eds., *Essays and Reviews*.
7 Altholz, *Anatomy of a Controversy*, 64 (on pamphlets); and Shea and Whitla, eds., *Essays and Reviews*, 42–3 (on edited collections).
8 Chadwick, *The Secularization of the European Mind in the Nineteenth Century*, 191.
9 It should be noted, however, that the term "Broad Church" gained prominence after its appearance in an anonymous article: see [W.J. Conybeare], "Church Parties, Past and Present," *Edinburgh Review* (1853): 272–342; and Ellis, *Seven against Christ*, 2–3.

10 For a history of the Broach Church movement, see Jones, *The Broad Church*.
11 Benjamin Jowett, "On the Interpretation of Scripture," in *Essays and Reviews*, 2nd edn (London: Parker, 1860), 338, 375. See also Stevens, *The Historical Jesus and the Literary Imagination, 1860–1920*, 16.
12 Shea and Whitla, eds., *Essays and Reviews*, 5.
13 [Frederic Harrison], "Neo-Christianity," *Westminster Review* 18:2 (1860): 293–332.
14 *Essays and Reviews*, 2nd edn (London: Parker, 1860), preface.
15 [Harrison], "Neo-Christianity," 293.
16 Ibid., 330.
17 On this episode, see Hesketh, *Of Apes and Ancestors*.
18 [Samuel Wilberforce], "Essays and Reviews," *Quarterly Review* 109 (January 1861): 256.
19 On Temple in particular, see Hesketh, "Frederick Temple and the *Essays and Reviews* Controversy"; and Shea and Whitla, eds., *Essays and Reviews*, 46–54.
20 Ibid., 648.
21 Quoted in W.R. Fremantle, "Oxford Essays and Reviews [Episcopal Manifesto]," *The Times*, 12 February 1861, 10.
22 Lightman, *Victorian Popularizers of Science*, 34.
23 J.W. Colenso, *The Pentateuch and Book of Joshua Critically Examined*, 2nd edn (London: Longman, Green, Longman, Roberts & Green, 1862), vi–vii. However, see Larsen, "Bishop Colenso and His Critics," 436–40, for a discussion of the moral critique of the Bible that informed Colenso's biblical criticism.
24 Colenso, *The Pentateuch and Book of Joshua Critically Examined*, viii–ix.
25 Ibid., x.
26 Ibid., xvii.
27 Shea and Whitla, eds., *Essays and Reviews*, 849.
28 Colenso, *The Pentateuch and Book of Joshua Critically Examined*, xx.
29 Ibid., xxiv.
30 Ibid., xxvi.
31 Ibid., xxxv.
32 Shea and Whitla, eds., *Essays and Reviews*, 848.
33 Ibid., 853.
34 Quoted ibid., 854.
35 For a full text of the synodical condemnation, see ibid., 672–6. See also Altholz, *Anatomy of a Controversy*, 123–4.
36 See, for instance, [Wilberforce], "Essays and Reviews," 306.
37 On the criticisms directed at Colenso by Broad Churchmen and liberal-minded clergymen, see Larsen, "Bishop Colenso and His Critics," 450–6.

38 Matthew Arnold, "Dr. Stanley's Lectures on the Jewish Church," *Macmillan's Magazine* 7:40 (February 1863): 332.
39 Especially useful on the strategies of theological publishing is Ledger-Lomas, "Mass Markets: Religion," and see esp. 355–6 for a discussion of Arnold, Colenso, and *Essays and Reviews*.
40 J. Llewelyn Davies, *St. Paul and Modern Thought: Remarks on Some of the Views Advanced in Professor Jowett's Commentary on St. Paul* (Cambridge: Macmillan, 1856), 79.
41 Ibid., 6.
42 Arnold, "Dr. Stanley's Lectures on the Jewish Church," 330.
43 Ibid., 331.
44 Ibid., 332.
45 [A.P. Stanley], "Essays and Reviews," *Edinburgh Review* 113 (April 1861): 461–500.
46 [Henry Sidgwick] A Cambridge Graduate, "Essays and Reviews," *The Times*, 20 February 1862, 12.
47 Rothblatt, *The Revolution of the Dons*, 134.
48 See von Arx, "The Victorian Crisis of Faith as a Crisis of Vocation"; and Lightman, *Origins of Agnosticism*, 101, 205.
49 Quoted in Benson, *The Life of Edward White Benson, sometime Archbishop of Canterbury*, 1:249–50.
50 Rothblatt, *The Revolution of the Dons*, 157.
51 Howsam, *Cheap Bibles*, xiv.
52 Cowling, *Religion and Public Doctrine in Modern England*, vol. 3: *Accommodations*, 87.

2 By the Author of *Essays on the Church*

1 Fyfe, *Science and Salvation*, 4.
2 There has been very little secondary work done on R.B. Seeley, but see Howsam, "Seeley, Robert Benton (1789–1886)."
3 Howsam, *Cheap Bibles*, 154.
4 Brake and Demoor, eds., *Dictionary of Nineteenth-Century Journalism in Great Britain and Ireland*, 114; see also Gleadle, "Charlotte Elizabeth Tonna and the Mobilization of Tory Women in Early Victorian England."
5 Lewis, *Lighten Their Darkness*, 154.
6 For the list of proprietors, see "Address," *Publishers' Circular* 1 (September 1837–December 1838), iv.
7 On the role faith played in the everyday lives of evangelicals, see Hilton, *The Age of Atonement*. See also the introduction to Fyfe, *Science and Salvation*.

8 Quoted in Balleine, *A History of the Evangelical Party in the Church of England*, 139. See also the "history" of the Church Pastoral Aid Society on its website.
9 Toon, *Evangelical Theology 1833–1856*, 44.
10 *Essays on the Church; with some reference to Mr. James's Work, entitled "Dissent and the Church of England." Reprinted with Additions from "The Christian Guardian"* (London: Seeley and Burnside, 1833).
11 See, for instance, "Essays on the Church," *Christian Remembrancer* 15:9 (September 1833), 538.
12 E.B. Pusey, *A Letter to the right rev. Father of God, Richard, Lord Bishop of Oxford, on the Tendency to Romanism Imputed to Doctrines held of old, as now, in the English Church* (1839; 4th edn, London: J.H. Parker, 1840).
13 By a Layman [R.B. Seeley], *Essays on the Church* (London: Seeley and Burnside, 1840), vii.
14 Ibid., xii. Seeley was here referring to Froude's *Remains of the Late Reverend Richard Hurrell Froude*, ed. John Henry Newman and John Keble, 4 vols. (London: J.G. & F. Rivington, 1838).
15 [Seeley], *Essays on the Church*, xii.
16 *Essays on Romanism* (London: Seeley and Burnside, 1839); *The Church of Christ in the Middle Ages: An Historical Sketch* (London: Seeley, Burnside, and Seeley, 1845); *Essays on the Bible* (London: Seeley, Jackson, & Halliday, 1870); *England's Training: An Historical Sketch* (London: Seeley and Co., 1885); *Is the Bible True? Seven Dialogues between James White and Richard Owen concerning the "Essays and Reviews" by the author of "Essays on the Church"* (London: Seeley, Jackson, & Halliday, 1862); *Is the Bible True? Familiar Dialogues between James White and Edward Owen concerning Bishop Colenso and the Pentateuch and the Testimony of Geology to the Bible* (London: Seeley, Jackson, & Halliday, 1863).
17 On the function of "by the author of ..." in anonymous publishing, see Griffin, "Anonymity and Authorship"; and Patten, "Anon."
18 See, for instance, "The Late Robert Benton Seeley," *Publishers' Circular*, 15 June 1886, 601–2; and "Mr. Seeley," *Athenaeum*, 5 June 1886, 748.
19 *The Octogenarian, A Grandfather's Address* (1853; 2nd edn, London: Nisbet and Co., 1855), 21.
20 *Church Missionary Review* 36 (1885): 120.
21 Griffin summarizes this narrative of publishing in "Anonymity and Authorship," 878–80.
22 Patten, "Anon."
23 And, of course, even Dickens regularly relied on pseudonyms such as "Boz," which itself became a particular identity. Readers moreover expected that Boz's works would be written in a particular style. See Patten, *Charles Dickens and "Boz."*

24 Griffin, "Anonymity and Authorship," 880.
25 Patten, "Anon."
26 This certainly follows from Fyfe's argument that most evangelicals writing on science did not seek to reject new discoveries but to situate them properly with their faith. See Fyfe, *Science and Salvation.*
27 [R.B. Seeley], *Is the Bible True? Familiar Dialogues between James White and Edward Owen concerning Bishop Colenso and the Pentateuch and the Testimony of Geology to the Bible,* 80–1.
28 Ibid., 80.
29 Ibid., 82.
30 Ibid., 78.
31 Ibid., 79.
32 Ibid., 89.
33 Ibid., 120.
34 Ibid., 124.
35 Ibid., 127.
36 Ibid., 129.
37 Ibid., 121.
38 Ibid., 136.
39 Ibid., 138.
40 [R.B. Seeley], *Is the Bible True? Seven Dialogues between James White and Richard Owen concerning the "Essays and Reviews" by the author of "Essays on the Church,"* 2.
41 Ibid., 3.
42 Ibid.
43 Ibid., 4.
44 Ibid., 10.
45 Ibid., 12.
46 Ibid., 17–18.
47 Ibid., 28.
48 Ibid., 47.
49 Ibid., 110.
50 Ibid., 111.
51 Ibid., 112–13.
52 Ibid., 118.
53 [R.B. Seeley], *Is the Bible True? Familiar Dialogues between James White and Edward Owen concerning Bishop Colenso and the Pentateuch and the Testimony of Geology to the Bible,* 2.
54 Ibid., 3.
55 Ibid., 7.

56 Ibid., 11.
57 Ibid., 24.
58 Ibid., 38.
59 Ibid., 59.
60 Ibid., 69.

3 Father and Son

1 Fyfe, *Science and Salvation*, 21.
2 See, for instance, Turner, *Contesting Cultural Authority*, ch. 3.
3 Edmund Gosse, *Father and Son: A Study of Two Temperaments* (London: W. Heinemann, 1907).
4 Philip Gosse, *Omphalos: An Attempt to Untie the Geological Knot* (London: John Van Voorst, 1857).
5 For brief considerations of Seeley's relationship with his father, see Hilton, *The Age of Atonement*, 334; Shannon, "John Robert Seeley and the Idea of a National Church"; Soffer, "History and Religion," 135–6; and Wormell, *Sir John Seeley and the Uses of History*, 12–15.
6 John Seeley to R.B. Seeley, 8 September 185?, Seeley Papers, MS 309/2A/2.
7 Wormell makes this entirely justifiable assumption in *Sir John Seeley and the Uses of History*, 14.
8 Reardon, "Maurice, (John) Frederick Denison (1805–1872)."
9 See, for instance, John Octavius Johnston, *Life and Letters of Henry Parry Liddon* (London: Longmans, Green, and Co., 1904), 72–5; and Morris, *F.D. Maurice and the Crisis of Christian Authority*, 161.
10 This is paraphrased from Shakespeare, *As You Like It* (1623), 5.4.103–6.
11 John Seeley to R.B. Seeley, 29 September 185?, Seeley Papers, MS 309/2A/2.
12 John Seeley to Mary Seeley, 3 April 1859, Seeley Papers, MS 309/2A/3.
13 On the genre of the historical Jesus, see Schweitzer, *The Quest of the Historical Jesus*. For the historical Jesus in Victorian Britain, see Stevens, *The Historical Jesus and the Literary Imagination, 1860–1920*; and Pals, *The Victorian "Lives" of Jesus*.
14 See Pals, *The Victorian "Lives" of Jesus*, 26.
15 *Westminster and Foreign Quarterly Review* 47 (April 1847): 138. See Stevens, *The Historical Jesus and the Literary Imagination, 1860–1920*, 38–9, for an analysis of this review as well as the Victorian reception of Strauss more broadly.
16 Stevens, *The Historical Jesus and the Literary Imagination, 1860–1920*, 39.
17 Ibid.
18 Ibid., 33.
19 Ibid., 34.

20 Jones, *The Broad Church*, 3; and Morgan, *The House of Macmillan (1843–1943)*, 35.
21 John Tulloch, *The Christ of the Gospels and the Christ of Modern Criticism: Lectures on M. Renan's "Vie de Jésus"* (London: Macmillan, 1864), 3–4.
22 F.D. Maurice, "Christmas Thoughts on Renan's Vie de Jésus," *Macmillan's Magazine* 9:51 (1864): 197.
23 On Seeley's positivist influences, see Wormell, *Sir John Seeley and the Uses of History*, 31–2.
24 The next few paragraphs on positivism and theology are indebted to Cashdollar, *Transformation of Theology, 1830–1890*; and Wright, *The Religion of Humanity*.
25 *North British Review* 15 (August 1851): 311; quoted in Cashdollar, *Transformation of Theology, 1830–1890*, 46.
26 Thomas Henry Huxley, *Lay Sermons, Addresses, and Reviews* (New York: Appleton, 1870), 140.
27 Ibid., 161.
28 On the intersection of Christian socialism and positivism in Seeley's life, see Wormell, *Sir John Robert Seeley*, 16–22.

4 The Victorian Jesus

1 Owen Chadwick makes a similar claim, stating that the "title misled … The book was neither a biography nor a portrait of a man. It was a study of the foundations of Christian morality." Chadwick, *The Victorian Church, Part II*, 64.
2 *Ecce Homo: A Survey in the Life and Work of Jesus Christ*, 5th edn (London: Macmillan, 1866), 18. That Seeley would refer to the Gospels as "Christ's biographies" goes some way to indicate just how remote he was from a critical perspective.
3 Ibid., xxi.
4 Ibid., xxii.
5 Ibid., xxii.
6 Ibid., 1–2.
7 Ibid., 2.
8 Ibid., 3.
9 Ibid., 4.
10 Ibid., 12.
11 Ibid., 15.
12 Ibid., 19.
13 Ibid., 20.
14 Ibid., 27.

15 Ibid., 36.
16 Ibid., 34.
17 Ibid., 35.
18 Ibid., 39–40.
19 Ibid., 37.
20 Ibid., 40.
21 Ibid., 43.
22 Ibid., 42–3.
23 Ibid., 43–4.
24 Ibid., 48.
25 Ibid., 51.
26 Ibid., 69.
27 Ibid., 75.
28 Ibid., 79.
29 Ibid., 123–4.
30 Ibid., 125–6.
31 Ibid., 129.
32 Ibid., 130.
33 Ibid., 131.
34 Ibid., 140.
35 Ibid., 156.
36 Ibid., 164.
37 Ibid., 156.
38 Ibid., 165.
39 Ibid., 180.
40 Ibid., 186.
41 Ibid., 188.
42 Ibid., 160–1.
43 Ibid., 200.
44 Ibid., 200–2.
45 Ibid., 202–3.
46 Ibid., 299.
47 Ibid., 170.
48 Ibid., 322.

5 A Dangerous Book

1 [R.W. Church], "Ecce Homo," *Guardian*, 7 February 1866, 139, republished in Church, "Ecce Homo," in *Occasional Papers* (London: Macmillan, 1897), 133. All references will be to the original version.

2 *Reader*, 25 November 1865, 608. See also *Spectator*, 25 November 1865, 1323; *Examiner*, 25 November 1865, 755; and *Athenaeum*, 25 November 1865, 716.
3 Morgan, *The House of Macmillan (1843–1943)*, 20.
4 VanArsdel, "Macmillan Family (*per. c.*1840–1986)."
5 On this specifically, see Morgan, *The House of Macmillan (1843–1943)*, 34–6.
6 Alexander Macmillan to James Burn, 18 June 1861, in Charles L. Graves, *Life and Letters of Alexander Macmillan* (London: Macmillan, 1910), 175.
7 Morgan, *The House of Macmillan (1843–1943)*, 64.
8 John Seeley to Alexander Macmillan, 14 August 1882, Macmillan Archive, Add MS 55074: 22. I must thank Leslie Howsam for sharing her transcription of this letter with me.
9 Wormell, *Sir John Seeley and the Uses of History*, 16.
10 Seeley to Alexander Macmillan, 14 August 1882, Macmillan Archive, Add MS 55074: 22. "I do not suppose you remember that you found me playing at cricket on Parker's Piece in the summer of 1857 & told me that you had just come from seeing him [Daniel] die!"
11 VanArsdel, "Macmillan Family (*per. c.*1840–1986)." See also Morgan, *The House of Macmillan (1843–1943)*, 50–3.
12 Macmillan to Seeley, 13 December 1865, Seeley Papers, MS 903/3A/1.
13 Macmillan to Seeley [no date], Seeley Papers, MS 903/3A/1.
14 [R.H. Hutton], "Ecce Homo," *Spectator*, 23 December 1865, 1436. The "Neander" referenced in this quote is August Neander, who wrote *Das Leben Jesu Christi, in seinem geschichilichen Zusammenhang und seiner geschichtlichen Entwickelung* (1837).
15 Ibid., 1436.
16 Ibid., 1437 ("defect of method"), 1438 ("and his readers").
17 Macmillan to Seeley, 23 December 1865, Seeley Papers, MS 903/3A/1.
18 "To the Editor of the 'Spectator,'" *Spectator*, 30 December 1865, 1467.
19 See the letter from R.H. Hutton, 27 December 1865, Seeley Papers, MS 903/3A/2. Interestingly, even though Hutton claimed that he had nothing left to say about *Ecce Homo*, he would, according to Macmillan (letter to Seeley, 3 April 1866, Seeley Papers, MS 903/3A/1), pen another review that appeared in the *North British Review*, March 1866.
20 "Ecce Homo," *Patriot*, 28 December 1865, 843.
21 Ibid., 844.
22 Macmillan to Seeley, 1 January 1866, Seeley Papers, MS 903/3A/1.
23 "Ecce Homo," *British Quarterly Review* 43:85 (January 1866): 230.
24 Ibid.
25 Ibid., 232.
26 Macmillan to Seeley, 4 January 1866, Seeley Papers, MS 903/3A/1.

27 "Ecce Homo," *Reader*, 16 January 1866, 33.
28 "Ecce Homo," *Church and State Review*, 2 February 1866, 73, in Seeley Papers, MS 903/3B/2.
29 Macmillan to Seeley, 13 February 1866, Seeley Papers, MS 903/3A/1.
30 Macmillan to Seeley, 21 February 1866, Seeley Papers, MS 903/3A/1.
31 27 February 1866, Macmillan letter book, Macmillan Archive, Add MS 55385(2): 540 (letterbook).
32 W.E. Gladstone to Macmillan, 25 December 1865, Seeley Papers, MS 903/3A/1.
33 On Gladstone as a Tractarian, see Lynch, "Was Gladstone a Tractarian?"
34 Gladstone to Macmillan, 25 December 1865, Seeley Papers, MS 903/3A/1.
35 Macmillan to Seeley, 29 December 1865, Seeley Papers, MS 903/3A/1.
36 Macmillan to Gladstone, 29 December 1865, Gladstone Papers, British Library, Add MS 44127: 4–5. A copy of this letter was enclosed in Macmillan to Seeley, 1 January 1866, Seeley Papers, MS 903/3A/1.
37 On the *Guardian*, see Brake and Demoor, eds., *Dictionary of Nineteenth-Century Journalism in Great Britain and Ireland*, 263.
38 Murphy, "Church, Richard William (1815–1890)."
39 Gladstone to Macmillan, 31 December 1865, Seeley Papers, MS 903/3A/2.
40 Macmillan to Seeley [undated, though sent sometime after 9 January 1866], Seeley Papers, MS 903/3A/1. The letter to Gladstone is also discussed in Macmillan to Seeley, 9 January 1866, Seeley Papers, MS 903/3A/1.
41 Seeley to Gladstone, [no date, draft], Seeley Papers, MS 903/3A/1.
42 Macmillan to Seeley, 8 February 1866, Seeley Papers, MS 903/3A/1.
43 [Church], "Ecce Homo," 139.
44 Ibid.
45 Ibid., 140.
46 Ibid., 141.
47 Macmillan to Seeley, 8 February 1866, Seeley Papers, MS 903/3A/1.
48 Macmillan to Seeley, 13 February 1866, Seeley Papers, MS 903/3A/1.
49 Matthew Arnold to Mary Penrose Arnold, 23 February 1866, in *The Letters of Matthew Arnold*, vol. 3: *1866–1870*, ed. Cecil Y. Lang (Charlottesville and London: University Press of Virginia, 1998), 15.
50 Macmillan to Seeley, 8 February 1866, Seeley Papers, MS 903/3A/1.
51 Macmillan to Seeley, 3 February 1886, Seeley Papers, MS 903/3A/1.

6 Vomited from the Jaws of Hell

1 *Reader*, 10 February 1866, 163.
2 "Ecce Homo," *John Bull*, 24 February 1866, 132.
3 "Ecce Homo," *Record*, 14 March 1866, 4.

4 Goldwin Smith to Macmillan, 27 December 1865, Seeley Papers, MS 309/3A/1.
5 Macmillan to Seeley, 29 December 1865, Seeley Papers, MS 309/3A/1.
6 F.D. Maurice to Macmillan, 4 January 1866, Seeley Papers, MS 309/3A/1.
7 Macmillan to Seeley, 4 January 1866, Seeley Papers, MS 309/3A/1.
8 Macmillan to Seeley, 21 February 1866, Seeley Papers, MS 309/3A/1. See Stanley to Macmillan, 17 February 1866, Seeley Papers, MS 309/3A/1.
9 Dean of St Paul's to Macmillan, 9 January 1866, Seeley Papers, MS 309/3A/1.
10 Spencer R. Drummond to Macmillan (for the author of Ecce Homo), 14 March 1866, Seeley Papers, Add MS 309/3A/1.
11 Quoted in Macmillan to Seeley, 29 March 1866, MS 903/3A/1; and see Rose Kingsley to Macmillan, 7 March 1866, Seeley Papers, MS 903/3A/2.
12 Henry Sidgwick to Macmillan, [February 1866], Seeley Papers, MS 309/3A/1.
13 Macmillan to Seeley, 21 February 1866, Seeley Papers, MS 309/3A/1.
14 Macmillan to Seeley, 3 February 1866, Seeley Papers, MS 309/3A/1.
15 Macmillan to Seeley, 15 March 1866, Seeley Papers, MS 309/3A/1. See also, for instance, Macmillan to Seeley, 26 March 1866, Seeley Papers, MS 309/3A/1; and Macmillan to Seeley, 29 March 1866, Seeley Papers, MS 309/3A/1.
16 See, for instance, *Athenaeum*, 14 April 1866, 510; *Spectator*, 14 April 1866, 421.
17 Macmillan to Seeley, 29 March 1866, Seeley Papers, MS 309/3A/1.
18 Rigg, "Denison, George Anthony (1805–1896)."
19 George A. Denison, "Ecce Homo," *Churchman*, 29 March 1866, 304. A clipping of the article has been preserved in Seeley Papers, MS 903/3B/1.
20 "Latest News," *John Bull*, 31 March 1866, 224.
21 Macmillan to Seeley, 3 April 1866, Seeley Papers, MS 309/3A/1.
22 [Whitwell Elwin], "Ecce Homo," *Quarterly Review* 119:238 (April 1866): 529.
23 See, for instance, *Athenaeum*, 5 May 1866, 610.
24 "The Quarterly Review," *Reader*, 14 April 1866, 369.
25 "The Quarterlies," *John Bull*, 14 April 1866, 241.
26 Macmillan to Seeley, 10 April 1866, Seeley Papers, MS 309/3A/1.
27 W.F. Pollock to Macmillan, 16 April 1866, Seeley Papers, MS 309/3A/1.
28 Macmillan to Seeley, 10 April 1866, Seeley Papers, MS 309/3A/1.
29 W., "Two Views of 'Ecce Homo,'" *Journal of Sacred Literature and Biblical Record* 8:18 (July 1866): 353n.
30 *Spectator*, 14 April 1866, 421.
31 This was reported in "Literary Gossip," *London Review*, 12 May 1866, 545.
32 *Spectator*, 21 April 1866, 451.
33 *Spectator*, 12 May 1866, 533.
34 See, for instance, ads in *Athenaeum*, 5 May 1866, 610; *Spectator*, 5 May 1866, 506; *Examiner*, 19 May 1866, 320; and *Observer*, 20 May 1866, 1.

35 Edward T. Vaughan, "Ecce Homo," *Contemporary Review* 2 (May 1866): 40–58.
36 As quoted in *John Bull*, 12 May 1866, 318.
37 See, for instance, "London Correspondence," *Bury and Norwich Post, and Suffolk Herald*, 8 May 1866, 8; "Earl Shaftesbury and 'Ecce Homo,'" *Daily News*, 9 May 1866; "Earl Shaftesbury and 'Ecce Homo,'" *Liverpool Mercury*, 9 May 1866; "Summary of This Morning's News," *Pall Mall Gazette*, 9 May 1866; "News of the Day," *Birmingham Daily Post*, 10 May 1866; "Earl Shaftesbury and 'Ecce Homo,'" *Hull Packet and East Riding Times*, 11 May 1866; "Earl Shaftesbury and 'Ecce Homo,'" *Derby Mercury*, 16 May 1866; "Earl Shaftesbury and 'Ecce Homo,'" *Manchester Guardian*, 28 May 1866.
38 See "Lord Shaftesbury's Metaphor," *Pall Mall Gazette*, 12 May 1866; and "Lord Shaftesbury's Inferno," *Saturday Review*, 19 May 1866, 586–7. "Lord Shaftesbury's Inferno" was also reprinted in the *Observer*, 22 May 1866, 7.
39 "Lord Shaftesbury's Inferno," *Saturday Review*, 586–7.
40 Peter Bayne, "Ecce Homo," *Fortnightly Review* 5:26 (June 1866): 131.
41 Lord Shaftesbury, diary entry, 12 May 1865, Shaftesbury Papers, MS 62/SHA/PD/8.
42 "Summary of This Morning's News," *Pall Mall Gazette*, 25 October 1866; "Lord Shaftesbury on Church Questions," *Morning Post*, 25 October 1866, 2; "Miscellaneous News," *Preston Guardian*, 27 October 1866.
43 *Spectator*, 21 April 1866, 451.
44 *Spectator*, 12 May 1866, 533.
45 *Spectator*, 9 June 1866, 647.
46 For full references to these reviews, see the section on "Reviews of *Ecce Homo*" in the Bibliography.
47 "Ecce Homo," *London Review*, 12 May 1866, 538.
48 Ibid., 539.
49 Vaughan, "Ecce Homo," 43–4.
50 Ibid., 43.
51 Ibid., 53.
52 "Ecce Homo," *Quiver* 1:38 (June 1866): 598.

7 A Sheep in Wolf's Clothing

1 Major Bell to Mary Elizabeth Roberts, 8 May 1866, Buckle Papers, Post 1650 MS 66 n. 6, letter 7.
2 [James Fitzjames Stephen], "Ecce Homo. First Notice," *Fraser's Magazine* 73:438 (June 1866): 746.

3 Ibid., 747–8.
4 Ibid., 752.
5 Ibid., 753.
6 Wormell, *Sir John Seeley and the Uses of History*, 16.
7 Henry Sidgwick to his mother, 29 January [1866], in Arthur Sidgwick and Eleanor Mildred Sidgwick, eds., *Henry Sidgwick: A Memoir* (London: Macmillan, 1906), 143.
8 Sidgwick to H.G. Dakyns, 4 May 1866, ibid., 147–8.
9 Sidgwick to Roden Noel, no date, ibid., 150.
10 Sidgwick to Seeley, 7 May 1866, Seeley Papers, MS 903/3A/3.
11 [Henry Sidgwick], "Ecce Homo," *Westminster Review* 30:1 (July 1866): 87. For a useful summary of Sidgwick's critique of *Ecce Homo*, see Schneewind, *Sidgwick's Ethics and Victorian Moral Philosophy*, 28–35.
12 [Sidgwick], "Ecce Homo," 58.
13 Ibid., 87.
14 Ibid., 59.
15 Ibid., 61.
16 Ibid., 68.
17 Ibid., 70.
18 Ibid., 68.
19 Sidgwick to Seeley, 9 May 1866 (first letter), Seeley Papers, MS 903/3A/3.
20 Sidgwick to Seeley, 9 May 1866 (second letter), Seeley Papers, MS 903/3A/3.
21 James, *Henry Sidgwick*, 19.
22 Macmillan to Seeley, 20 April 1866, Seeley Papers, MS 309/3A/1.
23 Macmillan to Seeley, 4 June 1866, Seeley Papers, MS 309/3A/1.
24 Ibid.
25 [J.R. Seeley], *Ecce Homo: A Survey in the Life and Work of Jesus Christ*, 5th edn (London: Macmillan, 1866), vi.
26 Ibid., vii.
27 Ibid., vii–ix, on ix.
28 Ibid., x.
29 Ibid., xi.
30 Ibid, xiii.
31 Ibid., xvi.
32 Ibid., xvii–xviii.
33 Ibid., xviii.
34 Ibid., xix.

35 Ibid., xx.
36 "The New Preface to *Ecce Homo*," *Spectator*, 14 July 1866, 769–70.
37 [John Henry Newman], "Ecce Homo," *Month*, 4 June 1866, 564.
38 Ibid., 555.
39 Ibid., 555–6.
40 Ibid., 557.
41 Ibid., 562.
42 See, for instance, [Sidgwick], "Ecce Homo," 69: "There is no more fruitful source of error in history than the determination to find the tree in the seed, and to attribute to the originators of important social changes detailed foresight as to the shape those changes were to assume."
43 [Newman], "Ecce Homo," 562–3.
44 Ibid., 564.
45 Ibid., 565.
46 Ibid., 565.
47 Macmillan to Seeley, 31 May 1866, Seeley Papers, MS 309/3A/1.

8 Shrewd Conjecture

1 W.E. Gladstone, *On Ecce Homo* (London: Strahan & Co., 1868), 1.
2 "Literary Gossip," *London Review*, 19 May 1866, 572. Also quoted in *Manchester Guardian*, 22 May 1866, 3.
3 "Metropolitan Gossip," *Sheffield and Rotherham Independent*, 26 May 1866, 6.
4 Ibid.
5 "Literary Gossip," *London Review*, 19 May 1866, 572; also quoted in *Manchester Guardian*, 22 May 1866, 3.
6 "General News," *Glasgow Herald*, 11 August 1866; "This Evening's News," *Pall Mall Gazette*, 24 May 1866; "Spirit of the Press," *Bradford Observer*, 24 May 1866; and "Literary, Scientific, and Art," *Birmingham Daily Post*, 21 May 1866.
7 As reported in "General News," *Glasgow Herald*, 11 August 1866.
8 "Athenaea," *John Bull*, 19 May 1866, 342. "The Oxford Times denies the report that Mr. George Waring is the author of 'Ecce Homo.'"
9 *Oxford Times* quoted in *Manchester Guardian*, 15 May 1866, 5.
10 "This Evening's News," *Pall Mall Gazette*, 3 May 1866.
11 See "Literary Gossip," *London Review*, 5 May 1866, 517; and "Literary, Scientific, and Art," *Birmingham Daily Post*, 7 May 1866.
12 "This Evening's News," *Pall Mall Gazette*, 28 June 1866. The *Star*'s suggestion was reprinted in "The Author of 'Ecce Homo,'" *Glasgow Herald*, 29 June 1866.

13 *Friend of India*, 9 August 1866, 941.
14 *Australasian*, 1 December 1866, 1093.
15 Quoted in "General News," *Nottinghamshire Guardian*, 21 September 1866; and "General Intelligence," *Berrow's Worcester Journal*, 22 September 1866, 6.
16 *Melbourne Punch*, 30 August 1866, 72.
17 As suggested by E.A. Freeman to Macmillan, 20 May 1866, Macmillan Archive, Add MS 55049:50.
18 I write about how *Nemesis of Faith* played a role in the reception of Froude's later histories in "Fanatical Hatred or Brotherly Love?"
19 According to Nicoll, when "*Ecce Homo* was published, Dean Church thought it was written by Newman. He wrote his magnificent essay on the book under this conviction." *The Church's One Foundation*, 57n96.
20 Macmillan to Seeley, 3 April 1866, Seeley Papers, MS 903/3A/1.
21 Matthew Arnold to Mary Penrose Arnold, 23 February 1866, in *The Letters of Matthew Arnold*, vol. 3: *1866–1870*, ed. Cecil Y. Lang (Charlottesville and London: University Press of Virginia, 1998), 15.
22 The *Bookseller*'s "shrewd conjecture" was quoted in "Art and Literary Gossip," *Manchester Times*, 1 September 1866; and "General News," *Dundee Courier & Argus*, 11 September 1866. It was also briefly reported in "Latest Intelligence," *Freeman's Journal and Daily Commercial Advertiser*, 18 September 1866; and "Literary, Scientific, and Art," *Birmingham Daily Post*, 27 August 1866.
23 "Ecce Homo," *Patriot*, 19 July 1866, 475. This piece is also referred to in "Local and General," *Leeds Mercury*, 14 November 1866.
24 "Brief Literary Notices," *London Quarterly Review* 26:51 (April 1866): 257.
25 "London Correspondence," *Trewman's Exeter Flying Post or Plymouth and Cornish Advertiser*, 16 May 1866.
26 "The Authorship of 'Ecce Homo,'" *Friend of India*, 9 August 1866, 948.
27 [A.P. Stanley] APS, "Ecce Homo," *Macmillan's Magazine* 14:8 (June 1866): 134.
28 Ibid.
29 Macmillan to Seeley, 10 April 1866, Seeley Papers, MS 903/3A/1.
30 Stanley to Macmillan, 17 February 1866, Seeley Papers, MS 309/3A/1.
31 "Literature," *Dundee Courier & Argus*, 11 June 1866; "Authorship of 'Ecce Homo,'" *Sheffield and Rotherham Independent*, 11 June 1866, 4; "Authorship of 'Ecce Homo,'" *York Herald*, 9 June 1866, 2; "Authorship of 'Ecce Homo,'" *Glasgow Herald*, 23 June 1866; and "Authorship of 'Ecce Homo,'" *Bury and Norwich Post, and Suffolk Herald*, 26 June 1866, 8.
32 "Ecce Homo," *Fortnightly Review* 5:26 (June 1866): 129.
33 Ibid., 131.
34 Ibid., 137.

35 George Warington, *"Ecce Homo" and Its Detractors: A Review* (London: William Skeffington, 1866), 34, from Seeley Papers, MS 903/3B/1.
36 Promotional pamphlet for *Ecce Homo: A Survey in the Life and Work of Jesus Christ* (Boston: Roberts Brothers, 1866), 3, from Seeley Papers MS 903/3C/11.
37 Quoted in "Literary Gossip," *London Review*, 14 July 1866, 5.
38 "Ecce Homo," *Christian Remembrancer* 52:133 (July 1866): 130.
39 Ibid., 152.
40 Ibid., 153.

9 White Lies

1 [A.P. Stanley] APS, "Ecce Homo," *Macmillan's Magazine* 14:8 (June 1866): 134.
2 Morgan, *The House of Macmillan (1843–1943)*, 79.
3 Ibid., 135.
4 Freeman to Macmillan, 20 May 1866, Macmillan Archive, Add MS 55049: 50; and see Macmillan to Freeman, 14 May 1866, Macmillan Archive, Add MS 55386(1): 53.
5 Arnold to Macmillan, 16 April 1866, Macmillan Archive, Add MS 54978: 30. See also *The Letters of Matthew Arnold*, vol. 3: *1866–1870*, ed. Cecil Y. Lang (Charlottesville and London: University Press of Virginia, 1998), 39; and Jones, *The Broad Church*, 285.
6 Macmillan to Seeley, 13 February 1866, Seeley Papers, MS 903/3A/1.
7 Shairp to Macmillan, 8 February 1866, Seeley Papers, MS 903/3A/1.
8 Shairp's biographer unfortunately did not realize that Macmillan's letter to Shairp was a ruse. He therefore reproduced the fiction that Messrs Macmillan "were as puzzled as others to know who could be the author, and they had come at length to guess that Mr. Shairp might be the man." Knight, *Principal Shairp and His Friends*, 292. This mistake has been reproduced in Jones's *The Broad Church*, 285.
9 Shairp to Macmillan, 8 February 1866, Seeley Papers, MS 903/3A/1.
10 Macmillan to Seeley, 25 June 1866, Seeley Papers, MS 903/3A/1.
11 Macmillan to unknown, 22 February 1866, Macmillan Archive, Add MS 55385(2): 522.
12 Macmillan to Seeley, 13 December 1865, Seeley Papers, MS 903/3A/1.
13 Macmillan to Seeley, 23 December 1865, Seeley Papers, MS 903/3A/1.
14 Rose Kingsley to Macmillan, 7 March 1866, Seeley Papers, MS 903/3A/1.
15 *Oxford Times* quoted in *Manchester Guardian*, 15 May 1866, 5.
16 Seeley to Sidgwick, 15 May [1866], Sidgwick Papers, Add MS c. 95/67 (1–2); and George Warington, *"Ecce Homo" and Its Detractors: A Review* (London: William Skeffington, 1866).

17 Macmillan to Seeley, 22 May 1866, Seeley Papers, MS 903/3A/1.
18 Macmillan to Seeley, 31 May 1866, Seeley Papers, MS 903/3A/1.
19 Macmillan to Seeley, 21 February 1866, Seeley Papers, MS 903/3A/1.
20 Macmillan to Seeley, 29 March 1866; and Macmillan to Seeley, 3 April 1866, Seeley Papers, MS 903/3A/1.
21 Sidgwick to Seeley, 9 May [1866] (first letter), Seeley Papers, MS 903/3A/3.
22 Ibid.
23 Seeley to Henry Sidgwick, 9 May [1866], Sidgwick Papers, Add MS c.95/64(1).
24 Seeley to Sidgwick, 15 May [1866], Sidgwick Papers, Add MS c. 95/67 (1–2).
25 Edwin A. Abbott to Seeley, 5 April 1866, Seeley Papers, MS 903/3A/4. See also Macmillan to Seeley, 18 April 1866, Seeley Papers, MS 903/3A/1. "How does the Abbott matter stand. I don't think if he can be made to hold water there will be much difficulty in other quarters."
26 Macmillan to Seeley, 13 June 1866, Seeley Papers, MS 903/3A/1.
27 On this shift away from anonymous publishing in Victorian journalism, see Maurer, "Anonymity vs. Signature in Victorian Reviewing," and Gowan Dawson, Richard Noakes, and Jonathan R. Topham, "Introduction," in Cantor et al., *Science in the Nineteenth-Century Periodical*, 20–1. On the rise of a liberal individualist identity in connection with the debates about anonymous journalism, see Hadley, *Living Liberalism*, ch. 3.
28 On the central role of morality in Victorian intellectual life, see Collini, *Public Moralists*.
29 W., "Two Views of 'Ecce Homo,'" *Journal of Sacred Literature and Biblical Record* (July 1866): 334.
30 John Henry Newman, *Apologia pro Vita Sua: An Authoritative Basic Text of the Newman-Kingsley Controversy, Origin and Reception of the Apologia Essays in Criticism*, ed. David J. DeLaura (New York: W.W. Norton and Company, 1968).
31 "White Lies," *Saturday Review*, 5 May 1866, 526.
32 Seeley to Henry Sidgwick, 9 May [1866], Sidgwick Papers, Add MS c.95/64(1).
33 J.R. Seeley to Henry Allon, unknown date, in *Letters to a Victorian Editor*, ed. Albert Peel (London: Independent Press, 1929), 221. Peel suggests that the date of this letter is 9 December 1865, but this cannot be accurate, as Napoleon III had not been suggested as an author then, nor had there been much, if any, speculation about *Ecce Homo*'s authorship at that point. More likely it was written between June and July 1866. Unfortunately, I have been unable to track down the original letter. Peel does refer to the letter, however, in Albert Peel to Frances Seeley, 17 January 1929 and 22 January 1929, Seeley Papers, MS 903B/1/1.

34 "Ecce Homo," *Patriot*, 19 July 1866, 475.
35 Macmillan to Seeley, 25 June 1866, Seeley Papers, MS 903/3A/1.
36 J. Venn, "Personal Reminiscences of J.R. Seeley," *Caian* (1895): 164, Seeley Papers, MS 903/3A/2.
37 Edwin A. Abbott to Seeley, 5 April 1866, Seeley Papers, MS 903/3A/4.
38 Davies to Macmillan, 13 November 1866, Seeley Papers, MS 903/3A/2.
39 "*Ecce Homo!*" *Spectator*, 10 November 1866, 1243.
40 "Literary Gossip," *London Review*, 17 November 1866, 560; "Literary Gossip," *Athenaeum*, 17 November 1866, 647; "Miscellanea," *Reader*, 17 November 1866, 945; "Athenaea," *John Bull*, 17 November 1866, 744.
41 "Miscellanea," *Reader*, 17 November 1866, 945.
42 "This Evening's News," *Pall Mall Gazette*, 12 November 1866; "A Publisher's Secret," *Bury and Norwich Post*, 13 November 1866, 3; "Local and General," *Leeds Mercury*, 14 November 1866; "General Intelligence," *Bradford Observer*, 15 November 1866, 3; "The Week," *Ipswich Journal*, 17 November 1866; "The Cotton Manufacture," *Leicester Chronicle and the Leicestershire Mercury*, 17 November 1866, 6; and "Ecclesiastical and University," *Berrow's Worcester Journal*, 17 November 1866, 7 (note that in *Berrow's Worcester Journal*, Seeley's name was misspelled as Selwyn).
43 "Topics of the Week," *Era*, 18 November 1866.
44 Macmillan to Seeley, 12 December 1866, Seeley Papers, MS 903/3A/1.

10 Behold the Man

1 Macmillan to Davies, 13 November 1866, Macmillan Archive, Add MS 55842: 58.
2 Macmillan to Seeley, 12 December 1866, Seeley Papers, MS 903/3A/1.
3 Davies to Macmillan, 13 November 1866, Seeley Papers, MS 903/3A/2.
4 "Our Weekly Gossip," *Athenaeum*, 2 December 1854.
5 Secord, *Victorian Sensation*, 395–6.
6 Ibid., 369.
7 Rose Kingsley to Macmillan, 7 March 1866, Seeley Papers, MS 903/3A/1.
8 On this point, see Mullan, *Anonymity*, 287.
9 Macmillan to Gladstone, 29 December 1865, Gladstone Papers, Add MS 44246: 4–5. The copy was enclosed in Macmillan to Seeley, 29 December 1865, Seeley Papers, MS 903/3A/1.
10 Macmillan to Seeley, 22 May 1866, Seeley Papers MSS 903/3A/1, emphasis added.
11 Stanley to Macmillan, 17 February 1866, Seeley Papers, MS 903/3A/1.
12 Sidgwick to Seeley, 15 November 1866, Seeley Papers, MS 903/3A/3.

13 Macmillan to Seeley, 12 December 1866, Seeley Papers, MS 903/3A/1.
14 W.H. Parry to Seeley, 15 January 1867, Seeley Papers, MS 903/3A/4.
15 J. Skelton to Seeley, Seeley Papers, 24 December 1866, MS 903/3A/4.
16 Frederick Fitch to Seeley, 16 November 1866, Seeley Papers, MS 903/1B/18.
17 Gibbins, "Mayor, Joseph Bickersteth (1828–1916)."
18 Sidgwick to Seeley, 9 May [1866] (first letter), Seeley Papers, MS 903/3A/3.
19 J.B. Mayor to Seeley, 25 November 1866, Seeley Papers, MS 903/3A/4.
20 Seeley to Mayor, 26 November [1866], Seeley Papers, MS 903/1A/1.
21 Ibid.
22 Maria Seeley to Seeley, Seeley Papers, MS 903/3A/4.
23 Seeley to Anne Seeley, 25 November [1866?], Seeley Papers, MS 903/3A/4.
24 Wormell, *Sir John Seeley and the Uses of History*, 15.
25 J.B. Mayor to Seeley, 25 November 1866, Seeley Papers, MS 903/3A/4.
26 Macmillan to Gladstone, 25 June 1867, Gladstone Papers, Add MS 44127: 8.
27 Morely, *The Life of William Ewart Gladstone*, 166. "The mask of the anonymous had much to do, he [Gladstone] thought, with its popularity, as had happened to the *Vestiges of Creation*. Undoubtedly when the mask fell off, interest dropped."
28 See Macmillan to Seeley, 15 January 1867; 18 March 1867; and 15 March 1867, Seeley Papers, MS 903/3A/1.
29 W.E. Gladstone, *On Ecce Homo* (London: Strahan & Co., 1868).
30 See, for instance, "Mr. Gladstone on *Ecce Homo*," *Saturday Review*, 8 February 1868, 168; and "Ecce Homo," *London Review*, 22 August 1868, 247.
31 Macmillan to Gladstone, 2 January 1868, Gladstone Papers, Add MS 44127: 11.
32 J.R. Seeley, *Livy Books I–X, with Introduction, Historical Examination, and Notes* (Oxford: Clarendon Press, 1871).
33 Macmillan was the publisher for Oxford University Press from 1863 until 1881.
34 Macmillan to Seeley, 18 April 1866, Seeley Papers, MS 903/3A/1.
35 Macmillan to Seeley, 25 June 1866; Macmillan to Seeley 12 October 1866, Seeley Papers, MS 903/3A/1.
36 Macmillan to Seeley, 27 March 1867, Seeley Papers, MS 903/3A/1.
37 J.R. Seeley, *A Paper Read Before the University College Students' Christian Association* (London: H.K. Lewis, 1867), 1, from Seeley Papers, MS 903/4/2.
38 Ibid., 2.
39 Ibid., 3.
40 Ibid., 4.
41 Ibid., 6–7.
42 Ibid., 7.

43 Ibid., 9.
44 J.R. Seeley, "A Letter Written by Professor J.R. Seeley, M.A., on March 23rd, 1868," in William Scarell Lean, *Are the Churches in Touch with the People* (York: John Sampson; London: Simpkin, Marshall, Hamilton, Kent, & Co., 1895), 15, from Seeley Papers, MS 903/4/2.
45 Ibid., 16.
46 See, for instance, "Essays on Church Policy," *Saturday Review*, 15 August 1868, 235–7; "Essays on Church Policy," *Christian Remembrancer* 56:142 (October 1868): 356–80; "Essays on Church Policy," *Spectator*, 27 June 1868, 771; and "Essays on Church Policy," *British Quarterly Review* 48:95 (July 1868): 285–90.
47 J.R. Seeley, "The Church as a Teacher of Morality," in W.L. Clay, *Essays on Church Policy* (London: Macmillan, 1868), 248–9.
48 Ibid., 251.
49 Ibid., 253.
50 Ibid., 253–4.
51 Ibid., 257.
52 Ibid., 266.
53 "Essays on Church Policy," *Saturday Review*, 15 August 1868, 235.
54 "Essays on Church Policy," *Christian Remembrancer* 56:142 (October 1868): 376.
55 Ibid., 377.
56 "Essays on Church Policy," *Saturday Review*, 15 August 1868, 236.
57 "Essays on Church Policy," *Spectator*, 27 June 1868, 771.

11 Behold the Historian

1 See Gladstone's letter book, 30 March 1869, Gladstone Papers, Add MS 44536: 137; 20 August 1869, Gladstone Papers, Add MS 44537: 32.
2 Gladstone to Bishop of St Davids, 2 July 1869, Gladstone's letter book, Gladstone Papers, Add MS 44537: 2.
3 Gladstone's letter book, 21 July 1869, Add MS 44537:12 (Kingsley); 27 August 1869, Add MS 44537: 39 (Maurice).
4 J.R. Seeley to Charles Kingsley, 1869, Kingsley Papers, Add MS 41299: 142–5.
5 Gladstone to Seeley, 2 September 1869, Seeley Papers, MS 903/1B/23; Gladstone's letter book, 2 September 1869, Gladstone Papers, Add MS 44537: 46; Seeley to Gladstone, 14 September 1869, Gladstone Papers, Add MS 44422:33. On Seeley's appointment, see Hesketh, *The Science of History in Victorian Britain*, 74–8; Wormell, *Sir John Seeley and the Uses of History*, 42; and Chadwick, "Charles Kingsley at Cambridge," 317.

6 Quoted in Elton, *Return to Essentials*, 102.
7 "Modern History at Cambridge," *Saturday Review*, 19 February 1870, 241.
8 J.R. Seeley, "The Teaching of Politics: An Inaugural Lecture Delivered at Cambridge," *Lectures and Essays* (London: Macmillan, 1870), 317. For a useful summary of Seeley's lecture, see Wormell, *Sir John Seeley and the Uses of History*, 43–7.
9 J.R. Seeley, *Life and Times of Stein, or Germany and Prussia in the Napoleonic Age*, 3 vols. (Cambridge: Cambridge University Press, 1878), 1:vii–viii.
10 Ibid., 1:xvi.
11 "Life and Times of Stein," *Examiner*, 18 January 1879, 84.
12 Ibid.
13 Ibid.
14 "Seeley's Life and Times of Stein," *Saturday Review*, 7 January 1879, 146.
15 Ibid., 147.
16 Ibid., 147.
17 "The Reorganizer of Modern Germany: Stein," *Westminster Review* 55:2 (April 1879): 336.
18 George Strachey, "Life and Times of Stein," *Academy*, 1 February 1879, 88.
19 "The Life and Times of Stein," *Spectator*, 1 February 1879, 52–3.
20 "The Life and Times of Stein," *International Review* 6 (1879): 582.
21 "Biographies of the Season," *London Society* 35:208 (April 1879): 376.
22 Seeley to Robert Spence Watson, 6 June [1879?], Seeley Papers, MS 903A/1/3.
23 Seeley to Watson, 18 July [1880], Seeley Papers, MS 903A/1/3.
24 J.R. Seeley, "History and Politics," *Macmillan's Magazine*, 40 (1879): 289–99, 369–78, 449–58; 41 (1879): 23–32.
25 I compare Seeley's and Freeman's critiques of Macaulay in "Writing History in Macaulay's Shadow."
26 Seeley to C.E. Maurice, 8 April [1880/1? copy], Seeley Papers, MS 903/1A/2.
27 Seeley to Mary Seeley, 3 April 1859, Seeley Papers, MS 309/2A/3.
28 J. Llewelyn Davies to the editor of the *Spectator* [unpublished], 22 January 1895, Seeley Papers, MS 903A/1/7.
29 Seeley, *Life and Times of Stein, or Germany and Prussia in the Napoleonic Age*, 1:vii.
30 Ibid., 1:xvi.
31 Bell, "John Robert Seeley and the Political Theology of Empire," in *Remaking the World*, ch. 11.
32 Caroline Jebb, *Life and Letters of Sir Richard Claverhouse Jebb* (Cambridge: Cambridge University Press, 1907), 85–6.

12 Fulfilling a Promise

1 After Seeley became Regius Professor of Modern History at Cambridge, Macmillan published a volume of his lectures and essays: J.R. Seeley, *Lectures and Essays* (London: Macmillan, 1870).
2 On Macmillan's cultivation of various historians and historical projects, see Howsam, *Past into Print*.
3 *Ecce Homo: A Survey in the Life and Work of Jesus Christ* (Boston: Roberts Brothers, 1866).
4 "Ecce Homo," *Dublin University Magazine*, 68:403 (July 1866): 76–7.
5 [Joseph Parker], *Ecce Deus: Essays on the Life and Doctrine of Jesus Christ with Controversial Notes on "Ecce Homo* (Edinburgh: T & T Clark, 1867), viii.
6 Macmillan to Seeley, 25 June 1866, Seeley Papers, MS 903/3A/1.
7 Seeley to Macmillan, 25 June 1872, Macmillan Archive, Add MS 55074: 7.
8 Macmillan to Seeley, 2 July 1872, Macmillan Archive, Add MS 55392: 710.
9 "Natural Religion" was originally published in ten instalments: [J.R. Seeley], "Natural Religion," *Macmillan's Magazine* 31 (1875): 357–7, 473–83; 32 (1875): 193–204, 481–9; 33 (1875–6): 1–9, 385–93; 34 (1876): 156–67, 522–34; 35 (1877): 417–29; 37 (1878): 177–91. The first four instalments were republished in the United States by Edward Youmans under the title "Deeper Harmonies of Science and Religion," *Popular Science Monthly* 7 (1875): 66–79, 301–14, 573–87; 8 (1875), 225–6. It does not appear that any of the other instalments were reprinted, possibly because of their irregular appearance after that point. Many thanks to James Ungureanu for drawing my attention to these reprints.
10 Seeley to Macmillan, 7 July 1872, Macmillan Archive, Add MS 55074: 8.
11 *Macmillan's Magazine* 1 (November 1859).
12 Hadley, *Living Liberalism*, ch. 3. See also Worth, *Macmillan's Magazine, 1859–1907*, 32; Vincent, *The Culture of Secrecy*, 65–6; and Maurer, "Anonymity vs. Signature in Victorian Reviewing."
13 Thomas Hughes, "A Run in the World's Fair," *Macmillan's Magazine* 17:97 (November 1867): 86–94.
14 Alexander Macmillan to David Masson, 21 October 1867, Macmillan Archive, Add MS 54792: 90.
15 Macmillan to Masson, 22 October 1867, Macmillan Archive, Add MS 54792: 92.
16 J.R. Seeley, "English in Schools," *Macmillan's Magazine* 17:97 (November 1867): 75–86.
17 Macmillan to Masson, 23 October 1867, Macmillan Archive, Add MS 54792: 93.
18 Thomas Hughes, "Anonymous Journalism," *Macmillan's Magazine* 5:26 (December 1861): 160.
19 Ibid., 168.

20 Ibid., 159.
21 For Maurice's influence on the early years of *Macmillan's Magazine*, see Worth, *Macmillan's Magazine, 1859–1907*, 53–97.
22 F.D. Maurice to Charles Kingsley, 27 May 1858, in Frederick Maurice, ed., *The Life and Times of Frederick Denison Maurice*, 2 vols, 4th edn (London: Macmillan, 1885), 2:321–2.
23 Ibid., 2:322.
24 Ibid., 2:323.
25 Maurice to R.H. Hutton, 23 August 1858, ibid., 2:325.
26 Ibid., 2:324.
27 [E.S. Dallas], "Popular Literature," *Blackwood's Magazine* 85 (June 1859): 187. See also Hadley, *Living Liberalism*, 132.
28 Seeley to Macmillan, 1 February 1875, Macmillan Archive, Add MS 55074: 11.
29 "Ecce Homo," *Dublin University Magazine* 68:403 (July 1866): 76–7.
30 Seeley to George Grove, 13 December 1875, Macmillan Archive, Add MS 55074: 13.
31 Macmillan to Seeley, 11 January 1877, Macmillan Archive, Add MS 55401: 374; Grove to Seeley, 28 April 1877, Macmillan Archive, Add MS 55401: 434; Grove to Seeley, 25 June 1877, Macmillan Archive, Add MS 55401: 932; and Grove to Seeley, 6 October 1877, Seeley Papers, MS 903/1B/18: 18.
32 Macmillan to Seeley, 11 January 1877, Macmillan Archive, Add MS 55401: 374.
33 Craik to Seeley, 26 July 1877, Macmillan Archive, Add MS 55403: 274.
34 Seeley to Macmillan, 12 March [1877?], Macmillan Archive, Add MS 55049: 56.
35 Macmillan to Seeley, 30 January 1877, Macmillan Archive, Add MS 55401: 550.
36 Seeley to Macmillan, 6 February 1877, Macmillan Archive, Add MS 55074: 18.
37 Macmillan to Seeley, 20 June 1877, Macmillan Archive, Add MS 55403: 208.
38 Macmillan to Seeley, 11 December 1877, Macmillan Archive, Add MS 55404: 478.
39 Macmillan to Seeley, 1 January 1878, Macmillan Archive, Add MS 55404: 646.
40 See Macmillan to Seeley, 9 June 1879, Macmillan Archive, Add MS 55408(3): 1261; Macmillan to Seeley, 10 June 1879, Macmillan Archive, Add MS 55408(3): 1277; Grove to Seeley, 25 October 1879, Macmillan Archive, Add MS 55409(2): 806; and Macmillan to Seeley, 26 January 1880, Macmillan Archive, Add MS 55410(1): 112.
41 See, for instance, Macmillan to Seeley, 9 June 1879, Macmillan Archive, Add MS 55408(3): 1261; and Macmillan to Seeley, 26 January 1880, Macmillan Archive, Add MS 55410(1): 112.
42 Seeley to Macmillan, 28 January [1880], Macmillan Archive, Add MS 55074: 61.
43 Macmillan to Seeley, 24 May 1880, Macmillan Archive, Add MS 55409(2): 150.
44 See, for example, Macmillan to Seeley, 29 September 1881, Macmillan Archive, Add MS 55412(3): 1366.

13 By the Author of *Ecce Homo*

1 Macmillan to Seeley, 23 May 1882, Macmillan Archive, Add MS 55414(1): 178.
2 *Natural Religion, by the Author of "Ecce Homo"* (London: Macmillan, 1882), 263.
3 Ibid., v.
4 Ibid., v–vi.
5 Ibid., v.
6 Ibid.
7 The creation of the conflict thesis in the nineteenth century is often attributed to John William Draper, *History of the Conflict between Religion and Science* (New York: D. Appleton, 1874); and Andrew Dickson White, *A History of the Warfare of Science with Theology in Christendom*, 2 vols. (New York: D. Appleton, 1896). For a discussion of the origins of the conflict thesis, see the introduction in Numbers, ed., *Galileo Goes to Jail, and Other Myths about Science and Religion*.
8 See, for instance, Cowling, *Religion and Public Doctrine in Modern England*, vol. 3: *Accommodations*, 91.
9 *Ecce Homo: A Survey in the Life and Work of Jesus Christ*, 5th edn (London: Macmillan, 1866), 322.
10 Ibid., 323.
11 Seeley was more explicit about this in the original articles, referring to the "favourite conviction about the Unknown and Unknowable" among fashionable men of science. See [Seeley], "Natural Religion," *Macmillan's Magazine* 35:210 (April 1878): 424.
12 *Natural Religion, by the Author of "Ecce Homo,"* 172.
13 Ibid., 178.
14 Ibid., 26.
15 Ibid., 223.
16 Ibid., 271; Compare Seeley's discussion of prophecy in *Ecce Homo*, 4–10.
17 *Natural Religion, by the Author of "Ecce Homo,"* 243.
18 Ibid., 271–2.
19 Ibid., 296.
20 Cowling, *Religion and Public Doctrine in Modern England*, vol. 3: *Accommodations*, 91.
21 [W.T. Davidson], "Natural Religion," *Quarterly Review* 154 (October 1882): 447.
22 J. Robinson Gregory, "Natural Religion," *Wesleyan-Methodist Magazine* (December 1882): 914.
23 Ibid., 905.
24 "Natural Religion," *Athenaeum*, 29 July 1882, 136.
25 Ibid., 135.

26 J. Llewelyn Davies, "'Natural Religion,' by the Author of 'Ecce Homo,'" *Contemporary Review* 42 (September 1882): 454.
27 "Natural Religion," *Modern Review* 4 (January 1883): 45.
28 G.A. Simcox, "Natural Religion," *Nineteenth Century* 12 (September 1882): 392.
29 Ibid., 402.
30 On fears of heat-death, see, for instance, Myers, "Nineteenth-Century Popularizations of Thermodynamics and the Rhetoric of Social Prophecy."
31 Simcox, "Natural Religion," 393–4.
32 W.S. Lilly, "Natural Religion," *Dublin Review* 8 (October 1882), 491–2.
33 "The Author of 'Ecce Homo' on Natural Religion," *London Quarterly Review* 59 (October 1882): 161.
34 Gregory, "Natural Religion," 905.
35 "The Author of 'Ecce Homo' on Natural Religion," *Spectator*, 1 July 1882, 867.
36 J.R. Seeley, "Preface to the Second Edition," in *Natural Religion* (London: Macmillan, 1895), viii–ix.
37 [J.R. Seeley], "'Natural Religion,' and Its Drift," *Spectator*, 16 June 1883, 768.
38 Ibid.
39 Ibid., 767.
40 Stanley, *Huxley's Church and Maxwell's Demon.*
41 See Turner, *Between Science and Religion*, ch. 5.
42 Penman, "The History of the Word *Paranormal*."
43 F.W.H. Myers, "A New Eirenicon," *Fortnightly Review* 32:191 (November 1882): 607.
44 Though it should be noted that Seeley suggested to Macmillan that if *Natural Religion* were to be reviewed in *Macmillan's Magazine*, "Frederick Myers ought to be asked to write the notice." Seeley to Macmillan, n.d. [1882], Macmillan Archive, Add MS 55074: 22.
45 See, for instance, Oppenheim, *The Other Word.*
46 [Seeley], "'Natural Religion,' and Its Drift," 768.
47 Macmillan to Seeley, 20 November 1882, Macmillan Archive, Add MS 55414(3): 1436.
48 See the discussion about printing figures in Frederick Macmillan to Seeley, 11 July 1882, Macmillan Archive, Add MS 55414(1): 444 (1,430 copies); Frederick Macmillan to Seeley, 14 July 1882, Macmillan Archive, Add MS 55414(2): 530 (2,000 copies); Alexander Macmillan to Seeley, 24 July 1882, Macmillan Archive, Add MS 55414(2): 570 (discussion of the printing of a "third thousand"); Alexander Macmillan to Seeley, 3 August 1882, Macmillan Archive, Add MS 55414(2): 624 (decision to print the "third thousand" as the second edition); George Macmillan to Seeley,

30 August 1882, Macmillan Archive, Add MS 55414(2): 788 (subscription of the second edition); and Frederick Macmillan to Seeley, 7 December 1882, Macmillan Archive, Add MS 55415(1): 77 (issuing of the "fourth thousand").

49 Macmillan to Seeley, 29 January 1880, Macmillan Archive, Add MS 55409(2): 150; Macmillan to Roberts Brothers, 22 May 1882, Macmillan Archive, Add MS 55414(1): 161; and Macmillan to Seeley, 23 May 1882, Macmillan Archive, Add MS 55414(1): 178.
50 Frederick Macmillan to Seeley, 30 June 1890, Macmillan Archive, Add MS 55431(1): 95.
51 Seeley to Frederick Macmillan, June 1890, Macmillan Archive, Add MS 55074: 34. Frederick Macmillan suggested lowering the price to six shillings so that it could be reprinted in uniformity in both "size and price" with *Ecce Homo*: see Frederick Macmillan to Seeley, 7 June [1890], Macmillan Archive, Add MS 55430(1); and Frederick Macmillan to Seeley, 30 June [1890], Macmillan Archive, Add MS 55431(1): 95. Macmillan confirms this change in Frederick Macmillan to Seeley, 25 July [1890], Macmillan Archive, Add MS 55431(1): 367.
52 W.H. Simcox, "Natural Religion," *Academy*, 15 July 1882, 41; and Myers, "A New Eirenicon," 596.
53 "Natural Religion," *Athenaeum*, 136.
54 The same could not be said about the United States. R.W. Boodle, who reviewed *Natural Religion* in *Popular Science Monthly*, matter-of-factly stated that the author was "now universally recognized as Professor Seeley." See Boodle, "Natural Religion I," *Popular Science Monthly* 22 (February 1883): 515
55 Last Will and Testament of John Robert Seeley, 11 July 1872 (probate, 18 February 1895): Leeds Probate Registry, York Place, Leeds.
56 Mary Seeley to Macmillan Publishers, 11 May 1904, Macmillan Archive, Add MS 55074.
57 Macmillan and Co. to Mary Seeley, 16 May 1904, Macmillan Archive, Add MS 55476(2): 976.

14 Remembering the Author of *Ecce Homo*

1 Bell, *The Idea of Greater Britain*, ch. 6; and Bell, *Remaking the World*, ch. 11.
2 See Leslie Howsam's description of the publishing of *The Expansion of England* in "Imperial Publishers and the Idea of Colonial History, 1870–1916," 4–6.
3 It was, of course, said by Edward Freeman. See Hesketh, "'History Is Past Politics, Politics Present History.'"
4 Macmillan and Co. negotiated with Mary Seeley to have Seeley's *Ecce Homo*, *Natural Religion*, *The Expansion of England*, and *Lectures and Essays*

published as part of the Macmillan Eversley series in 1895. See Macmillan and Co. to Mary Seeley, 24 January 1895, Macmillan Archive, Add MS 55446(3): 1367; Macmillan and Co. to Mary Seeley, 28 January 1895, Macmillan Archive, Add MS 55446(3): 1422; Macmillan and Co. to Mary Seeley, 15 February 1895, Macmillan Archive, Add MS 55447(1): 303; and Macmillan and Co. to Mary Seeley, 20 February 1895, Macmillan Archive, Add MS 55447(1): 411.

5 "Professor Seeley," *National Observer*, 19 January 1895, 254.

6 Ibid., 255.

7 "Sir John Seeley," *National Review* 144 (February 1895): 730.

8 "Sir John Seeley," *Saturday Review*, 19 January 1895, 89.

9 "Sir John Seeley," *Manchester Guardian*, 15 January 1895, 4.

10 Maurice Todhunter, "Sir John Seeley," *Westminster Review* 145:5 (January 1896): 504.

11 "Sir John Seeley, K.C.M.G.," *Academy*, 19 January 1895, 57.

12 Joseph Jacobs, "Sir John Seeley," *Athenaeum*, 19 January 1895, 86.

13 "Sir John Seeley," *Saturday Review*, 89.

14 H.A.L. Fisher, "Sir John Seeley," *Fortnightly Review* 60:356 (August 1896): 185.

15 E.M.T., "A Literary Causerie," *Speaker*, 19 January 1895, 76.

16 Fisher, "Sir John Seeley," 185.

17 E.M.T., "A Literary Causerie," 76.

18 "Death of Sir John Seeley," *The Times*, 15 January 1895, 10.

19 "Death of Sir John R. Seeley," *New York Times*, 15 January 1895, 5.

20 "Sir John Seeley," *Saturday Review*, 89.

21 Joseph Jacobs, "Sir John Seeley," *Athenaeum*, 19 January 1895, 86–7.

22 *Spectator*, 19 January 1895, 71.

23 Todhunter, "Sir John Seeley," 503–4.

24 "Sir John Seeley," *Manchester Guardian*, 15 January 1895, 4.

25 Jacobs, "Sir John Seeley," 86–7.

26 *Spectator*, 19 January 1895, 71.

27 "Sir John Seeley," *National Review*, 731.

28 Alfred Church, "Professor Seeley's Religious Position," *Spectator*, 2 Febuary 1895, 162–3.

29 Strachey to Davies, 21 January 1895, Seeley Papers, MS 903A/1/7.

30 Davies to Lady Seeley, 24 January 1895, Seeley Papers, MS 903A/1/7.

31 Davies to the editor of the *Spectator* [unpublished], 22 January 1895, Seeley Papers, MS 903A/1/7.

32 Richmond Seeley to Mary Seeley, 18 February 1895, Seeley Papers, MS 903A/1/8.

33 The reasons for this are not entirely clear, though Wormell surmises that a lack of "life and letters" may be due to the fact that later generations

of historians would have been unfamiliar with Seeley's wide range of interests. Moreover, interest in Seeley was largely neglected by historians until the 1960s. Wormell, *Sir John Seeley and the Uses of History*, 1.

34 G.W. Prothero to Mary Seeley, 19 January 1895, Seeley Papers, MS 903A/1/5.

35 G.W. Prothero to Mary Seeley, 1 February 1895, Seeley Papers, MS 903A/1/5.

36 G.W. Prothero to Mary Seeley, 19 January 1895, Seeley Papers, MS 903A/1/5.

37 Richmond Seeley to Mary Seeley, 18 February 1895, Seeley Papers, MS 903A/1/8.

38 Prothero to Mary Seeley, 21 February 1895, Seeley Papers, MS 903A/1/5.

39 G.W. Prothero, "Memoir," in Seeley, *The Growth of British Policy: An Historical Essay*, vol. 1, 2nd edn (Cambridge: Cambridge University Press, 1897), x.

40 Ibid.

Epilogue: Anonymous Publishing and Universal History

1 A.A., "Alexander Macmillan," *Macmillan's Magazine* 73:437 (March 1896): 397–400; "Mr. A. Macmillan," *Athenaeum*, 1 February 1896, 150; and "We regret to record […]," *Spectator*, 1 February 1896, 154.

2 J.S.C., "Alexander Macmillan," *Academy*, 1 February 1896, 96.

3 Quoted in Charles L. Graves, *Life and Letters of Alexander Macmillan* (London: Macmillan, 1910), 253. Unfortunately, Graves does not provide any citation information for this quote, or any real context for it. I have also been unable to track down the original statement in any of Seeley's extant letters to Macmillan.

4 On the theological origins of universal history, see Megill, "'Big History' Old and New"; and Momigliano, "The Origins of Universal History."

5 *Natural Religion, by the Author of "Ecce Homo,"* 296.

6 Christian, "The Return of Universal History," 12.

7 Seeley to Kingsley, 1869, Kingsley Papers, Add MSS 41299: 144.

8 This follows from the recent work of Duncan Bell, which argues that Seeley's political histories need to be understood within the context of his theological writings. See, in particular, *Remaking the World*, ch. 11.

Bibliography

Manuscript Sources

Buckle Papers: Henry Thomas Buckle Collection, Post 1650 MS 66, University of Illinois at Urbana-Champaign.
Gladstone Papers: The Papers of W.E. Gladstone, Add MS 44086–44551, British Library.
Kingsley Papers: The Correspondence and Papers of Charles Kingsley, Add MS 41297–41299, British Library.
Macmillan Archive: The Correspondence and Papers of the Publishing Firm of Macmillan and Company, Add MS 54786–56035, British Library.
Seeley Papers: The Papers of John Robert Seeley, MS 903, Senate House Library, University of London.
Shaftesbury Papers: Lord Shaftesbury Papers, MS 62/SHA/PD/8, University of Southampton.
Sidgwick Papers: The Papers of Henry Sidgwick, Add MS. c.93–106, Trinity College, Cambridge.

Reviews of *Ecce Homo*

Ash, Edward. *"Ecce Homo," Its Character and Teaching*. London: W. Macintosh, 1868.
Bayne, Peter. "Ecce Homo." *Fortnightly Review* 5:26 (June 1866): 129–42.
Chudleigh, M. Ecclesia. *Thoughts on "Ecce Homo."* Truro: James R. Netherton, 1867, from Seeley Papers, MS 903/3B/1.
[Church, R.W.] "Ecce Homo." *Guardian*, 7 February 1866, 139–41. Republished in R.W. Church, "Ecce Homo," in *Occasional Papers*, 133–79. London: Macmillan, 1897.

The Credentials of Conscience. A Few Reasons for the Popularity of "Ecce Homo": and a Few Words about Christianity. London: Longmans, Green, Reader, and Dyer, 1868.
[Curteis, G.H.] "Strauss, Renan, and 'Ecce Homo.'" *Edinburgh Review* 124:254 (October 1866): 450–75.
Denison, George A. "Ecce Homo." *Churchman*, 29 March 1866, 304.
"Ecce Homo." *Achill Missionary Herald*, 14 August 1866, 345.
"Ecce Homo." *British Foreign Evangelical Review* 57 (July 1866): 557–70.
"Ecce Homo." *British Quarterly Review* 43:85 (January 1866): 229–32.
"Ecce Homo." *Christian Remembrancer* 52:133 (July 1866): 130–54.
"Ecce Homo." *Christian Observer* (July 1866): 498–520.
"Ecce Homo." *Church and State Review*, 2 February 1866, 73.
"Ecce Homo." *Dublin Review* (July 1866): 256.
"Ecce Homo." *Dublin University Magazine* 68:403 (July 1866): 75–91.
"Ecce Homo." *Examiner*, 26 May 1866, 323–5.
"Ecce Homo." *Fortnightly Review* 26 (June 1866): 129–42.
"Ecce Homo." *John Bull*, 24 February 1866, 132.
"Ecce Homo." *Literary Churchman*, 24 February 1866.
"Ecce Homo." *London Review*, 12 May 1866, 538–9.
"Ecce Homo." *North American Review* 103:1 (July 1866): 302–7.
"Ecce Homo." *Patriot*, 28 December 1865, 843–4.
"Ecce Homo." *Quiver* 1:38 (June 1866): 597–8.
"Ecce Homo." *Reader*, 16 January 1866, 33–4.
"Ecce Homo." *Record*, 14 March 1866, 4.
"Ecce Homo." *The Sword and the Trowel* (July 1866): 308–17.
Ecce Homo, A Critique by Presbyter Anglicanus, on Behalf of the Cause of Free Enquiry and Free Expression. West Cliff, Ramsgate: Thomas Scott, 1866, from Seeley Papers, MS 903/3A/1.
[Elwin, Whitwell]. "Ecce Homo." *Quarterly Review* 119:238 (April 1866): 515–29.
Gladstone, W.E. *On Ecce Homo*. London: Strahan & Co., 1868.
Glazebrook, Jas. K. *Ecce Homo: A Denial of the Peculiar Doctrines of Christianity, A Review*. Blackburn: John Alston, 1866.
Hood, Edwin Paxton. "Ecce Homo." *Eclectic Review* 10 (March 1866): 234–41.
[Hutton, R.H.] "Ecce Homo." *Spectator*, 23 December 1865, 1436–8.
[Hutton, R.H.] "Ecce Homo and Modern Scepticism." *North British Review* 44:87 (March 1866): 124–53.
"The Life and Work of Jesus Christ." *Christian World*, no date, in Seeley Papers, MS 903/3B/2.
"The Life and Work of Christ." *The Times*, 2 June 1868, 7.
MacColl, Malcolm. "'Ecce Homo' and Its Critics." *Review* (July 1866): 442–54, in Seeley Papers, MS 903/3B/2.

"The New Preface to *Ecce Homo*." *Spectator*, 14 July 1866, 769–70.
[Newman, John Henry]. "Ecce Homo." *Month*, 4 June 1866, 551–72.
Notes on "Ecce Homo." London: Hatchard and Co., 1866, from Seeley Papers, MS 903/3B/1. See also BL 4023.d.23.
[Parker, Joseph]. *Ecce Deus: Essays on the Life and Doctrine of Jesus Christ with Controversial Notes on "Ecce Homo."* Edinburgh: T & T Clark, 1867.
Passingham, Townshend. *A Protest against "Ecce Homo."* London: S.W. Partridge and Co., 1868, from BL RB.23 a.33810.
Perry, Charles. *"Ecce Homo": A Lecture delivered at Trinity Church, East Melbourne by the Right Rev. the Lord Bishop of Melbourne.* Melbourne: A.J. Smith, 1867.
Secretan, Rev. Samuel. *Remarks on "Ecce Homo."* Leicester: Wm. Penn Cox, n.d., from Seeley Papers, MS 903/3B/1.
[Sidgwick, Henry]. "Ecce Homo." *Westminster Review* 30:1 (July 1866): 58–88.
[Stanley, A.P.] APS. "Ecce Homo." *Macmillan's Magazine* 14:8 (June 1866): 134–42.
[Stephen, James Fitzjames]. "Ecce Homo. First Notice." *Fraser's Magazine* 73:438 (June 1866): 746–65.
[Stephen, James Fitzjames]. "Ecce Homo. Second Notice." *Fraser's Magazine* 74:439 (July 1866): 29–52.
[Thom, J.H.] J.H.T. "Ecce Homo." *Theological Review* 3:13 (April 1866): 161–87.
Vaughan, Edward T. "Ecce Homo." *Contemporary Review* 2 (May 1866): 40–58.
W. "Two Views of 'Ecce Homo.'" *Journal of Sacred Literature and Biblical Record* 8:18 (July 1866): 334–53.
Warington, George. *"Ecce Homo" and Its Detractors: A Review.* London: William Skeffington, 1866.

Other Printed Primary Sources

A.A. "Alexander Macmillan." *Macmillan's Magazine* 73:437 (March 1896): 397–400.
"Address." *Publishers' Circular* 1 (September 1837–December 1838): iii–iv.
Allon, Henry. *Letters to a Victorian Editor*. Ed. Albert Peel. London: Independent Press, 1929.
Arnold, Matthew. "Dr. Stanley's Lectures on the Jewish Church." *Macmillan's Magazine* 7:40 (February 1863): 327–36.
– *The Letters of Matthew Arnold*, vol. 3: *1866–1870*. Ed. Cecil Y. Lang. Charlottesville and London: University Press of Virginia, 1998.
"Art and Literary Gossip." *Manchester Times*, 1 September 1866.
"Athenaea." *John Bull*, 19 May 1866, 342.
"Athenaea." *John Bull*, 17 November 1866, 744.
Athenaeum. 14 April 1866, 510.
Athenaeum. 25 November 1865, 716.

Athenaeum. 5 May 1866, 610.
Australasian. 1 December 1866, 1093.
"Authorship of 'Ecce Homo.'" *Bury and Norwich Post, and Suffolk Herald*, 26 June 1866, 8.
"Authorship of 'Ecce Homo.'" *Glasgow Herald*, 23 June 1866.
"Authorship of 'Ecce Homo.'" *Sheffield and Rotherham Independent*, 11 June 1866, 4.
"Authorship of 'Ecce Homo.'" *York Herald*, 9 June 1866, 2.
"The Author of 'Ecce Homo.'" *Glasgow Herald*, 29 June 1866.
"The Author of 'Ecce Homo' on Natural Religion." *London Quarterly Review* 59 (October 1882): 161–92.
"The Author of 'Ecce Homo' on Natural Religion." *Spectator*, 1 July 1882, 865–7.
"The Authorship of 'Ecce Homo.'" *Friend of India*, 9 August 1866, 948.
"Biographies of the Season." *London Society* 35:208 (April 1879): 376.
Boodle, R.W. "Natural Religion I." *Popular Science Monthly* 22 (February 1883): 511–21.
"Brief Literary Notices." *London Quarterly Review* 26:51 (April 1866): 257.
Church, Alfred. "Professor Seeley's Religious Position." *Spectator*, 2 February 1895, 162–3.
Church Missionary Review 36 (1885): 120.
Colenso, J.W. *The Pentateuch and Book of Joshua Critically Examined*, 2nd edn. London: Longman, Green, Longman, Roberts & Green, 1862.
[Conybeare, W.J.] "Church Parties, Past and Present." *Edinburgh Review* (1853): 272–342.
"The Cotton Manufacture." *Leicester Chronicle and the Leicestershire Mercury*, 17 November 1866, 6.
[Dallas, E.S.] "Popular Literature." *Blackwood's Magazine* 85 (June 1859): 180–9.
[Davidson, W.T.] "Natural Religion." *Quarterly Review* 154 (October 1882): 425–47.
Davies, J. Llewelyn. *St. Paul and Modern Thought: Remarks on Some of the Views Advanced in Professor Jowett's Commentary on St. Paul*. Cambridge: Macmillan, 1856.
– "'Natural Religion,' by the Author of 'Ecce Homo.'" *Contemporary Review* 42 (September 1882): 442–54.
"Death of Sir John R. Seeley." *New York Times*, 15 January 1895, 5.
"Death of Sir John Seeley." *The Times*, 15 January 1895, 10.
Draper, John William. *History of the Conflict between Religion and Science*. New York: D. Appleton, 1874.
E.M.T. "A Literary Causerie." *Speaker*, 19 January 1895, 76–7.
"Earl Shaftesbury and 'Ecce Homo.'" *Daily News*, 9 May 1866.
"Earl Shaftesbury and 'Ecce Homo.'" *Derby Mercury*, 16 May 1866.

"Earl Shaftesbury and 'Ecce Homo.'" *Hull Packet and East Riding Times*, 11 May 1866.
"Earl Shaftesbury and 'Ecce Homo.'" *Liverpool Mercury*, 9 May 1866.
"Earl Shaftesbury and 'Ecce Homo.'" *Manchester Guardian*, 28 May 1866.
"Ecce Homo." *London Review*, 22 August 1868, 247–8.
"Ecce Homo." *Patriot*, 19 July 1866, 475.
"*Ecce Homo!*" *Spectator*, 10 November 1866, 1243.
"Ecclesiastical and University." *Berrow's Worcester Journal*, 17 November 1866, 7.
Essays and Reviews, 2nd edn. London: Parker, 1860.
"Essays on Church Policy." *British Quarterly Review* 48:95 (July 1868): 285–90.
"Essays on Church Policy." *Christian Remembrancer* 56:142 (October 1868): 356–80.
"Essays on Church Policy." *Saturday Review*, 15 August 1868, 235–7.
"Essays on Church Policy." *Spectator*, 27 June 1868, 770–1.
"Essays on the Church." *Christian Remembrancer* 15:9 (September 1833): 538.
Examiner. 19 May 1866, 320.
Examiner. 25 November 1865, 755.
Fisher, H.A.L. "Sir John Seeley." *Fortnightly Review* 60: 356 (August 1896): 183–98.
Fremantle, W.R. "Oxford Essays and Reviews [Episcopal Manifesto]." *The Times*, 12 February 1861, 10.
Friend of India. 9 August 1866, 941.
Froude, Richard Hurrell. *Remains of the Late Reverend Richard Hurrell Froude*. Ed. John Henry Newman and John Keble. 4 vols. London: J.G. & F. Rivington, 1838.
"General Intelligence." *Berrow's Worcester Journal*, 22 September 1866, 6.
"General Intelligence." *Bradford Observer*, 15 November 1866, 3.
"General News." *Dundee Courier & Argus*, 11 September 1866.
"General News." *Glasgow Herald*, 11 August 1866.
"General News." *Nottinghamshire Guardian*, 21 September 1866.
Gosse, Edmund. *Father and Son: A Study of Two Temperaments*. London: W. Heinemann, 1907.
Gosse, Philip. *Omphalos: An Attempt to Untie the Geological Knot*. London: John Van Voorst, 1857.
Gregory, J. Robinson. "Natural Religion." *Wesleyan-Methodist Magazine* (December 1882): 904–14.
Graves, Charles L. *Life and Letters of Alexander Macmillan*. London: Macmillan, 1910.
[Harrison, Frederic]. "Neo-Christianity." *Westminster Review* 8:2 (1860): 293–332.
Hughes, Thomas. "Anonymous Journalism." *Macmillan's Magazine* 5:26 (December 1861): 157–68.

– "A Run in the World's Fair." *Macmillan's Magazine* 17:97 (November 1867): 86–94.

Huxley, Thomas Henry. *Lay Sermons, Addresses, and Reviews*. New York: Appleton, 1870.

J.S.C. "Alexander Macmillan." *Academy*, 1 February 1896, 95–6.

Jacobs, Joseph. "Sir John Seeley." *Athenaeum*, 19 January 1895, 86–7.

Jebb, Caroline. *Life and Letters of Sir Richard Claverhouse Jebb*. Cambridge: Cambridge University Press, 1907.

John Bull, 12 May 1866, 318.

Johnston, John Octavius. *Life and Letters of Henry Parry Liddon*. London: Longmans, Green, and Co., 1904.

Jowett, Benjamin. "On the Interpretation of Scripture." In *Essays and Reviews*, 2nd edn. London: Parker, 1860.

"The Late Robert Benton Seeley." *Publishers' Circular*, 15 June 1886, 601–2.

"Latest Intelligence." *Freeman's Journal and Daily Commercial Advertiser*, 18 September 1866.

"Latest News." *John Bull*, 31 March 1866, 224.

"Life and Times of Stein." *Examiner*, 18 January 1879, 84–5.

"The Life and Times of Stein." *International Review* 6 (1879): 578–82.

"The Life and Times of Stein." *Spectator*, 1 February 1879, 52–3.

Lilly, W.S. "Natural Religion." *Dublin Review* 8 (October 1882): 491–3.

"Literary Gossip." *Athenaeum*, 17 November 1866, 647.

"Literary Gossip." *London Review*, 5 May 1866, 517.

"Literary Gossip." *London Review*, 12 May 1866, 545.

"Literary Gossip." *London Review*, 19 May 1866, 572.

"Literary Gossip." *London Review*, 14 July 1866, 5.

"Literary Gossip." *London Review*, 17 November 1866, 560.

"Literary, Scientific, and Art." *Birmingham Daily Post*, 7 May 1866.

"Literary, Scientific, and Art." *Birmingham Daily Post*, 21 May 1866.

"Literary, Scientific, and Art." *Birmingham Daily Post*, 27 August 1866.

"Literature." *Dundee Courier & Argus*, 11 June 1866.

"Local and General." *Leeds Mercury*, 14 November 1866.

"London Correspondence." *Bury and Norwich Post, and Suffolk Herald*, 8 May 1866, 8.

"London Correspondence." *Trewman's Exeter Flying Post or Plymouth and Cornish Advertiser*, 16 May 1866.

"Lord Shaftesbury on Church Questions." *Morning Post*, 25 October 1866, 2.

"Lord Shaftesbury's Inferno." *Saturday Review*, 19 May 1866, 586–7, also published in *Observer*, 22 May 1866, 7.

"Lord Shaftesbury's Metaphor." *Pall Mall Gazette*, 12 May 1866.

Macmillan's Magazine 1 (November 1859).
Manchester Guardian, 15 May 1866, 5.
Manchester Guardian, 22 May 1866, 3.
Maurice, F.D. "Christmas Thoughts on Renan's Vie de Jésus." *Macmillan's Magazine* 9:51 (1864): 190–7.
Maurice, Frederick, ed. *The Life and Times of Frederick Denison Maurice*. 2 vols, 4th edn. London: Macmillan, 1885.
Melbourne Punch, 30 August 1866, 72.
"Metropolitan Gossip." *Sheffield and Rotherham Independent*, 26 May 1866, 6.
"Miscellanea." *Reader*, 17 November 1866, 945.
"Miscellaneous News." *Preston Guardian*, 27 October 1866.
"Modern History at Cambridge." *Saturday Review*, 19 February 1870, 241–2.
"Mr. A. Macmillan." *Athenaeum*, 1 February 1896, 150.
"Mr. Gladstone on *Ecce Homo*." *Saturday Review*, 8 February 1868, 167–8.
"Mr. Seeley." *Athenaeum*, 5 June 1886, 748.
Myers, F.W.H. "A New Eirenicon." *Fortnightly Review* 32:191 (November 1882): 596–607.
"Natural Religion." *Athenaeum*, 29 July 1882, 135–6.
"Natural Religion." *Modern Review* 4 (January 1883): 24–51.
Newman, John Henry. *Apologia pro Vita Sua: An Authoritative Basic Text of the Newman-Kingsley Controversy, Origin and Reception of the Apologia Essays in Criticism*. Ed. David J. DeLaura. New York: W.W. Norton and Company, 1968.
"News of the Day." *Birmingham Daily Post*, 10 May 1866.
Observer, 20 May 1866, 1.
The Octogenarian, A Grandfather's Address. 1853; 2nd edn, London: Nisbet and Co., 1855.
"Our Weekly Gossip." *Athenaeum*, 2 December 1854.
[Parker, Joseph]. *Ecce Deus: Essays on the Life and Doctrine of Jesus Christ with Controversial Notes on "Ecce Homo*. Edinburgh: T & T Clark, 1867.
"A Popular Religious Work." *Bury and Norwich Post and Suffolk Herald*, 17 April 1866, 3.
"Professor Seeley." *National Observer*, 19 January 1895, 254–5.
Prothero, G.W. "Memoir." In J.R. Seeley, *The Growth of British Policy: An Historical Essay*. 2nd edn. Cambridge: Cambridge University Press, 1897.
"A Publisher's Secret." *Bury and Norwich Post*, 13 November 1866, 3.
Pusey, E.B. *A Letter to the right rev. Father of God, Richard, Lord Bishop of Oxford, on the Tendency to Romanism Imputed to Doctrines held of old, as now, in the English Church*. 1839; 4th edn, London: J.H. Parker, 1840.
"The Quarterlies." *John Bull*, 14 April 1866, 241.
"The Quarterly Review." *Reader*, 14 April 1866, 369.

Reader, 10 February 1866, 163.

Reader, 25 November 1865, 608.

"The Reorganizer of Modern Germany: Stein." *Westminster Review* 55:2 (April 1879): 329–59.

"Seeley's Life and Times of Stein." *Saturday Review*, 7 January 1879, 146–7.

[Seeley, J.R.] "To the Editor of the 'Spectator.'" *Spectator*, 30 December 1865, 1467.

[Seeley, J.R.] *Ecce Homo: A Survey in the Life and Work of Jesus Christ*. 5th edn. London: Macmillan, 1866.

[Seeley, J.R.] *Ecce Homo: A Survey in the Life and Work of Jesus Christ*. Boston: Roberts Brothers, 1866.

Seeley, J.R. *A Paper Read Before the University College Students' Christian Association*. London: H.K. Lewis, 1867.

Seeley, J.R. "English in Schools." *Macmillan's Magazine* 17:97 (November 1867): 75–86.

Seeley, J.R. "The Church as a Teacher of Morality." In *Essays on Church Policy*, ed. W.L. Clay, 247–91. London: Macmillan, 1868.

Seeley, J.R. *Lectures and Essays*. London: Macmillan, 1870.

Seeley, J.R. "The Teaching of Politics: An Inaugural Lecture Delivered at Cambridge. In *Lectures and Essays*, 290–317.

Seeley, J.R. *Livy Books 1–X, with Introduction, Historical Examination, and Notes*. Oxford: Clarendon Press, 1871.

[Seeley, J.R.] "Natural Religion." *Macmillan's Magazine* 31 (1875): 357–7, 473–83; 32 (1875): 193–204, 481–9; 33 (1875–6): 1–9, 385–93; 34 (1876): 156–67, 522–34; 35 (1877): 417–29; 37 (1878): 177–91.

[Seeley, J.R.] "Deeper Harmonies of Science and Religion." *Popular Science Monthly* 7 (1875): 66–79, 301–14, 573–87; 8 (1875): 225–6.

Seeley, J.R. *Life of Stein, or Germany and Prussia in the Napoleonic Age*. 3 vols. Cambridge: Cambridge University Press, 1878.

[Seeley, J.R.] *Natural Religion, by the Author of "Ecce Homo."* London: Macmillan, 1882.

[Seeley, J.R.] "'Natural Religion,' and Its Drift." *Spectator*, 16 June 1883, 767–8.

Seeley, J.R. *The Expansion of England: Two Courses of Lectures*. London: Macmillan, 1883.

Seeley, J.R. *Natural Religion*. London: Macmillan, 1895.

Seeley, J.R. "Preface to the Second Edition." In *Natural Religion*, v–ix.

Seeley, J.R. "A Letter Written by Professor J. R. Seeley, M.A., on March 23rd, 1868." In William Scarell Lean, *Are the Churches in Touch with the People*, 14–16. York: John Sampson; London: Simpkin, Marshall, Hamilton, Kent, & Co., 1895.

Seeley. J.R. *The Growth of British Policy: An Historical Essay*. Vol. 1, 2nd edn. Cambridge: Cambridge University Press, 1897.

[Seeley, R.B.] *Essays on the Church; with some reference to Mr. James's Work, entitled "Dissent and the Church of England." Reprinted with Additions from "The Christian Guardian."* London: Seeley and Burnside, 1833.

[Seeley, R.B.] *Essays on Romanism*. London: Seeley and Burnside, 1839.

[Seeley, R.B.] By a Layman. *Essays on the Church*. London: Seeley and Burnside, 1840.

[Seeley, R.B.] *The Church of Christ in the Middle Ages: An Historical Sketch*. London: Seeley, Burnside, and Seeley, 1845.

[Seeley, R.B.] *Is the Bible True? Seven Dialogues between James White and Richard Owen concerning the "Essays and Reviews" by the author of "Essays on the Church."* London: Seeley, Jackson, & Halliday, 1862.

[Seeley, R.B.] *Is the Bible True? Familiar Dialogues between James White and Edward Owen concerning Bishop Colenso and the Pentateuch and the Testimony of Geology to the Bible*. London: Seeley, Jackson, & Halliday, 1863.

[Seeley, R.B.] *Essays on the Bible*. London: Seeley, Jackson, & Halliday, 1870.

[Seeley, R.B.] *England's Training: An Historical Sketch*. London: Seeley and Co., 1885.

Sidgwick, Arthur, and Eleanor Mildred Sidgwick, eds. *Henry Sidgwick: A Memoir*. London: Macmillan, 1906.

[Sidgwick, Henry]. A Cambridge Graduate. "Essays and Reviews." *The Times*, 20 February 1862, 12.

Simcox, G.A. "Natural Religion." *Nineteenth Century* 12 (September 1882): 391–409.

Simcox, W.H. "Natural Religion." *Academy*, 15 July 1882, 41–2.

"Sir John Seeley." *Manchester Guardian*, 15 January 1895, 4.

"Sir John Seeley." *National Review* 144 (February 1895): 730–1.

"Sir John Seeley." *Saturday Review*, 19 January 1895, 89.

"Sir John Seeley, K.C.M.G." *Academy*, 19 January 1895, 57.

"Spirit of the Press." *Bradford Observer*, 24 May 1866.

Spectator, 25 November 1865, 1323.

Spectator, 12 May 1866, 533.

Spectator, 14 April 1866, 421.

Spectator, 21 April 1866, 451.

Spectator, 5 May 1866, 506.

Spectator, 9 June 1866, 647.

Spectator, 11 January 1868, 59.

Spectator, 19 January 1895, 71.

[Stanley, A.P.] "Essays and Reviews." *Edinburgh Review* 113 (April 1861): 461–500.

Strachey, George. "Life and Times of Stein." *Academy*, 1 February 1879, 88–90.

"Summary of This Morning's News." *Pall Mall Gazette*, 9 May 1866.

"Summary of This Morning's News." *Pall Mall Gazette*, 25 October 1866.

"This Evening's News." *Pall Mall Gazette*, 3 May 1866.

"This Evening's News." *Pall Mall Gazette*, 24 May 1866.
"This Evening's News." *Pall Mall Gazette*, 28 June 1866.
"This Evening's News." *Pall Mall Gazette*, 12 November 1866.
Todhunter, Maurice. "Sir John Seeley." *Westminster Review* 145:5 (January 1896): 503–9.
"Topics of the Week." *Era*, 18 November 1866.
Tulloch, John. *The Christ of the Gospels and the Christ of Modern Criticism: Lectures on M. Renan's "Vie de Jésus."* London: Macmillan, 1864.
Venn, J. "Personal Reminiscences of J.R. Seeley." *Caian* (1895): 164–70.
"We regret to record […]." *Spectator*, 1 February 1896, 154.
"The Week." *Ipswich Journal*, 17 November 1866.
Westminster and Foreign Quarterly Review 47 (April 1847): 136–74.
White, Andrew Dickson. *A History of the Warfare of Science with Theology in Christendom*. 2 vols. New York: D. Appleton, 1896.
"White Lies." *Saturday Review*, 5 May 1866, 525–6.
[Wilberforce, Samuel]. "Essays and Reviews." *Quarterly Review* 109 (January 1861): 248–306.

Secondary Sources

Adams, Thomas R., and Nicolas Barker. "A New Model for the Study of the Book." In *A Potencie of Life: Books in Society*, ed. Nicolas Barker, 5–43. London: British Library, 1993.
Altholz, Joseph L. *Anatomy of a Controversy: The Debate over "Essays and Reviews."* Aldershot: Scolar Press, 1994.
Balleine, G.R. *A History of the Evangelical Party in the Church of England*. London: Church Book Room Press, 1951.
Bell, Duncan. *The Idea of Greater Britain: Empire and the Future of World Order, 1860–1900*. Princeton: Princeton University Press, 2007.
– *Remaking the World: Essays on Liberalism and Empire*. Princeton: Princeton University Press, 2016.
Benn, Alfred William. *The History of English Rationalism in the Nineteenth Century*. 2 vols. London: Longmans Green, and Co., 1906.
Benson, Arthur Christopher. *The Life of Edward White Benson, sometime Archbishop of Canterbury*. Vol. 1. London: Macmillan, 1899.
Bowler, Peter. *The Non-Darwinian Revolution: Reinterpreting a Historical Myth*. Baltimore: Johns Hopkins University Press, 1988.
Brake, Laurel, and Marysa Demoor, eds. *Dictionary of Nineteenth-Century Journalism in Great Britain and Ireland*. Ghent: Academia Press; London: British Library, 2009.

Bremner, G.A., and Jonathan Conlin, eds. *Making History: Edward Augustus Freeman and Victorian Cultural Politics*. Oxford: Oxford University Press, 2015.

Burrow, J.W. "The Uses of Philology in Victorian Britain." In *Ideas and Institutions of Victorian Britain: Essays in Honour of George Kitson Clark*, ed. Robert Robson, 180–205. London: G. Bell & Sons, 1967.

Buurma, Rachel Sagner. "Anonymity, Corporate Authority, and the Archive: The Production of Authorship in Late-Victorian England." *Victorian Studies* 50:1 (2007): 15–42.

Cantor, Geoffrey, Gowan Dawson, Graeme Gooday, Richard Noakes, Sally Shuttleworth, and Jonathan R. Topham. *Science in the Nineteenth-Century Periodical: Reading the Magazine of Nature*. Cambridge: Cambridge University Press, 2004.

Cashdollar, Charles. *Transformation of Theology, 1830–1890: Positivism and Protestant Thought in Britain and America*. Princeton: Princeton University Press, 1989.

Chadwick, Owen. "Charles Kingsley at Cambridge." *Historical Journal* 18:2 (June 1975): 303–25.

– *The Secularization of the European Mind in the Nineteenth Century*. Cambridge: Cambridge University Press, 1975.

– *The Victorian Church, Part II*. New York: Oxford University Press, 1970.

Christian, David. "The Return of Universal History." *History and Theory* 49 (2010): 6–27.

Church Pastoral Aid Society. Website: http://www.cpas.org.uk/about-CPAS/history#.Ui0vUhaYWxI, accessed 24 September 2013.

Collini, Stefan. *Public Moralists: Political Thought and Intellectual Life in Britain 1850–1930*. Oxford: Oxford University Press, 1991.

Cowling, Maurice. *Religion and Public Doctrine in Modern England*. Vol. 3: *Accommodations*. Cambridge: Cambridge University Press, 2001.

Darnton, Robert. "What Is the History of Books?" *Daedalus* 111:3 (1982): 65–83.

Ellis, Ieuan. *Seven against Christ: A Study of "Essays and Reviews."* Leiden: E.J. Brill, 1980.

Elton, G.R. *Return to Essentials: Some Reflections on the Present State of Historical Study*. Cambridge: Cambridge University Press, 2002.

Ferris, Ina. *The Achievement of Literary Authority: Gender, History, and the Waverley Novels*. Ithaca: Cornell University Press, 1991.

Fyfe, Aileen. *Science and Salvation: Evangelical Popular Science Publishing in Victorian Britain*. Chicago and London: University of Chicago Press, 2004.

– *Steam-Powered Knowledge: William Chambers and the Business of Publishing, 1820–1860*. Chicago: University of Chicago Press, 2012.

Fyfe, Aileen, and Bernard Lightman, eds. *Science in the Marketplace: Nineteenth-Century Sites and Experiences*. Chicago: University of Chicago Press, 2007.

Gibbins, John R. "Mayor, Joseph Bickersteth (1828–1916)." *Oxford Dictionary of National Biography*. Oxford University Press, 2004; online edn, May 2009. http://www.oxforddnb.com.ezproxy.library.uq.edu.au/view/article/56421, accessed 24 January 2016.

Gleadle, Kathryn. "Charlotte Elizabeth Tonna and the Mobilization of Tory Women in Early Victorian England." *Historical Journal* 50:1 (2007): 97–117.

Griffin, Robert J. "Anonymity and Authorship." *New Literary History* 30:4 (1999): 877–95.

Griffin, Robert J., ed. *The Faces of Anonymity: Anonymous and Pseudonymous Publication from the Sixteenth to the Twentieth Century*. New York: Palgrave, 2003.

Hadley, Elaine. *Living Liberalism: Practical Citizenship in Mid-Victorian Britain*. Chicago: University of Chicago Press, 2010.

Hampton, David. *Evangelical Disenchantment: Nine Portraits of Faith and Doubt*. New Haven: Yale University Press, 2008.

Hesketh, Ian. "Fanatical Hatred or Brotherly Love? Rethinking E.A Freeman's Feud with J.A. Froude." In *Making History: Edward Augustus Freeman and Victorian Cultural Politics*, ed. Alex Bremner and Jonathan Conlin, 255–72. Oxford: Oxford University Press, 2015.

– "Frederick Temple and the *Essays and Reviews* Controversy." In *Cultural Olympians: Rugby School's Intellectual and Spiritual Leaders*, ed. Patrick Derham and John Taylor, 123–36. Buckingham: University of Buckingham Press, 2013.

– "'History Is Past Politics, Politics Present History': Who Said It?" *Notes and Queries* 61:1 (March 2014): 105–8.

– *Of Apes and Ancestors: Evolution, Christianity, and the Oxford Debate*. Toronto: University of Toronto Press, 2009.

– *The Science of History in Victorian Britain*. London: Pickering & Chatto, 2011.

– "Writing History in Macaulay's Shadow: J.R. Seeley, E.A. Freeman, and the Audience for Scientific History in Late Victorian Britain." *Journal of the Canadian Historical Association* 22:2 (2011): 30–56.

Hilton, Boyd. *The Age of Atonement: The Influence of Evangelicalism on Social and Economic Thought, 1795–1865*. Oxford: Clarendon Press, 1988.

Howsam, Leslie. "Academic Discipline or Literary Genre? The Establishment of Boundaries in Historical Writing." *Victorian Literature and Culture* (2004): 525–45.

– *Cheap Bibles: Nineteenth-Century Publishing and the British and Foreign Bible Society*. Cambridge: Cambridge University Press, 1991.

– "Imperial Publishers and the Idea of Colonial History, 1870–1916." *History of Intellectual Culture* 5:1 (2005): 1–15.
– *Kegan Paul, a Victorian Imprint: Publishers, Books and Cultural History*. Toronto: University of Toronto Press, 1998.
– *Old Books and New Histories: An Orientation to Studies in Book and Print Culture*. Toronto: University of Toronto Press, 2006.
– *Past into Print: The Publishing of History in Britain, 1850–1950*. Toronto: University of Toronto Press; London: British Library, 2009.
– "Seeley, Robert Benton (1789–1886)." *Oxford Dictionary of National Biography*. Oxford: Oxford University Press, 2004. http://www.oxforddnb.com.ezproxy.library.uq.edu.au/view/article/25027, accessed 8 September 2013.
James, D.G. *Henry Sidgwick: Science and Faith in Victorian England*. Oxford: Oxford University Press, 1970.
Johns, Adrian. *The Nature of the Book: Print and Knowledge in the Making*. Chicago: University of Chicago Press, 1998.
Jones, Tod E. *The Broad Church: A Biography of a Movement*. Lanham: Lexington Books, 2003.
Knight, William. *Principal Shairp and His Friends*. London: John Murray, 1888.
Koditschek, Theodore. *Liberalism, Imperialism, and the Historical Imagination: Nineteenth-Century Visions of a Greater Britain*. Cambridge: Cambridge University Press, 2011.
Larsen, Timothy. "Bishop Colenso and His Critics: The Strange Emergence of Biblical Criticism in Victorian Britain." *Scottish Journal of Theology* 50:4 (1997): 433–58.
Ledger-Lomas, Michael. "Mass Markets: Religion." In *The Cambridge History of the Book in Britain*, ed. David McKitterick, 324–58. Cambridge: Cambridge University Press, 2015.
Lewis, Donald M. *Lighten Their Darkness: The Evangelical Mission to Working-Class London, 1828–1860*. New York: Greenwood Press, 1986.
Liddle, Dallas. "Salesmen, Sportsmen, Mentors: Anonymity and Mid-Victorian Theories of Journalism." *Victorian Studies* 41:1 (Autumn 1997): 31–68.
Lightman, Bernard. "Darwin and the Popularization of Evolution." *Notes and Records of the Royal Society of London* 64:1 (2010): 5–24.
– *Origins of Agnosticism*. Baltimore: Johns Hopkins University Press, 1987.
– *Victorian Popularizers of Science: Designing Nature for New Audiences*. Chicago: University of Chicago Press, 2007.
Lynch, M.J. "Was Gladstone a Tractarian? W.E. Gladstone and the Oxford Movement, 1833–45." *Journal of Religious History* 8:2 (1975): 364–89.
Maurer, Oscar, Jr. "Anonymity vs. Signature in Victorian Reviewing." *Studies in English* 27 (June 1948): 1–27.

Megill, Allan. "'Big History' Old and New: Presuppositions, Limits, Alternatives." *Journal of the Philosophy of History* 9 (2015): 306–26.

Momigliano, Arnaldo. "The Origins of Universal History." In Momigliano, *On Pagans, Jews, and Christians*, 31–58. Middletown, CT: Wesleyan University Press, 1987.

Moore, James R. *The Post-Darwinian Controversies: A Study of the Protestant Struggle to Come to Terms with Darwin in Great Britain and America, 1870–1900*. Cambridge: Cambridge University Press, 1981.

Morely, John. *The Life of William Ewart Gladstone*. 1903; Cambridge: Cambridge University Press, 2011.

Morgan, Charles. *The House of Macmillan (1843–1943)*. London: Macmillan, 1943.

Morris, Jeremy. *F.D. Maurice and the Crisis of Christian Authority*. Oxford: Oxford University Press, 2005.

Mullan, John. *Anonymity: A Secret History*. London: Faber and Faber, 2007.

Murphy, G. Martin. "Church, Richard William (1815–1890)." *Oxford Dictionary of National Biography*. Oxford: Oxford University Press, 2004. http://www.oxforddnb.com.ezproxy.library.uq.edu.au/view/article/5389, accessed 13 December 2016.

Myers, Greg. "Nineteenth-Century Popularizations of Thermodynamics and the Rhetoric of Social Prophecy." *Victorian Studies* 29:1 (1985): 36–66.

Nicoll, W.R. *The Church's One Foundation: Christ and Recent Criticism*. New York: A.C. Armstrong, 1901.

Numbers, Ronald L., ed. *Galileo Goes to Jail, and Other Myths about Science and Religion*. Cambridge, MA: Harvard University Press, 2009.

Oppenheim, Janet. *The Other Word: Spiritualism and Psychical Research in England, 1850–1914*. Cambridge: Cambridge University Press, 1985.

Pals, Daniel L. "The Reception of 'Ecce Homo.'" *Historical Magazine of the Protestant Episcopal Church* 46:1 (1977): 63–84.

– *The Victorian "Lives" of Jesus*. San Antonio: Trinity University Press, 1982.

Patten, Robert L. "Anon." John Coffin Memorial Lecture in the History of the Book, School of Advanced Study, Institute of English Studies, University of London, 29 June 2006.

– *Charles Dickens and "Boz": The Birth of the Industrial-Age Author*. Cambridge: Cambridge University Press, 2012.

Penman, Leigh T.I. "The History of the Word *Paranormal*." *Notes & Queries* 62:1 (2012): 31–4.

Priest, Robert D. *The Gospel According to Renan: Reading, Writing, and Religion in Nineteenth-Century France*. Oxford: Oxford University Press, 2015.

Reardon, Bernard M.G. "Maurice, (John) Frederick Denison (1805–1872)." *Oxford Dictionary of National Biography*. Oxford: Oxford University Press,

2004. http://www.oxforddnb.com.ezproxy.library.uq.edu.au/view/article/18384, accessed 26 September 2013.

Rigg, J.M. "Denison, George Anthony (1805–1896)," rev. George Herring, *Oxford Dictionary of National Biography*. Oxford: Oxford University Press, 2004. http://www.oxforddnb.com/ezproxy.library.uq.edu.au/view/article/7488, accessed 27 January 2014.

Rothblatt, Sheldon. *The Revolution of the Dons: Cambridge and Society in Victorian England*. London: Faber and Faber, 1968.

Schneewind, J.B. *Sidgwick's Ethics and Victorian Moral Philosophy*. Oxford: Clarendon Press, 1977.

Schweitzer, Albert. *The Quest of the Historical Jesus*. 1906. Ed. John Bowden. London: SCM Press, 2000.

Secord, James. *Victorian Sensation: The Extraordinary Publication, Reception, Secret Authorship of "Vestiges of the Natural History of Creation."* Chicago: University of Chicago Press, 2000.

Shannon, R.T. "John Robert Seeley and the Idea of a National Church." In *Ideas and Institutions of Victorian Britain: Essays in Honour of George Kitson Clark*, ed. Robert Robson, 236–67. London: Bell, 1967.

Shea, Victor, and William Whitla, eds. *Essays and Reviews: The 1860 Text and Its Readings*. Charlottesville: University Press of Virginia, 2000.

Soffer, Reba N. "History and Religion: J.R. Seeley and the Burden of the Past." In *Religion and Irreligion in Victorian Society*, ed. R.W. Davis and R.J. Helmstadter, 133–50. New York: Routledge, 1992.

"Spanish Fresco Restoration Botched by Amateur." *BBC.com*, 23 August 2012. http://www.bbc.com/news/world-europe-19349921.

Stanley, Matthew. *Huxley's Church and Maxwell's Demon: From Theistic Science to Naturalistic Science*. Chicago: University of Chicago Press, 2014.

Stevens, Jennifer. *The Historical Jesus and the Literary Imagination, 1860–1920*. Liverpool: Liverpool University Press, 2010.

Toon, Peter. *Evangelical Theology 1833–1856: A Response to Tractarianism*. London: Marshall, Morgan & Scott, 1979.

Topham, Jonathan. "Beyond the 'Common Context': The Production and Reading of the Bridgewater Treatises." *Isis* 89 (1998): 233–62.

Turner, Frank M. *Between Science and Religion: The Reaction to Scientific Naturalism in Late Victorian England*. New Haven and London: Yale University Press, 1974.

– *Contesting Cultural Authority: Essays in Victorian Intellectual Life*. Cambridge: Cambridge University Press, 1993.

VanArsdel, Rosemary T. "Macmillan Family (*per. c.*1840–1986)." *Oxford Dictionary of National Biography*. Oxford: Oxford University Press, 2004.

online edn, May 2007. http://www.oxforddnb.com.ezproxy.library.uq.edu.au/view/article/63220, accessed 29 May 2015.

Vincent, David. *The Culture of Secrecy: Britain, 1832–1998*. Oxford: Oxford University Press, 1998.

von Arx, Jeffrey. "The Victorian Crisis of Faith as a Crisis of Vocation." In *Victorian Faith in Crisis: Essays on Continuity and Change in Nineteenth-Century Religious Belief*, ed. Bernard Lightman, 262–82. London: Macmillan, 1990.

White, Hayden. *Tropics of Discourse: Essays in Cultural Criticism*. Baltimore: Johns Hopkins University Press, 1978.

Wormell, Deborah. *Sir John Seeley and the Uses of History*. Cambridge: Cambridge University Press, 1980.

Worth, George J. *Macmillan's Magazine, 1859–1907: "No Flippancy or Abuse Allowed."* Burlington, VT: Ashgate, 2003.

Wright, T.R. *The Religion of Humanity: The Impact of Comtean Positivism on Victorian Britain*. Cambridge: Cambridge University Press, 1986.

Index

An italic *f* following a page reference indicates an illustration.

Studies in Book and Print Culture

General Editor: Leslie Howsam

Hazel Bell, *Indexers and Indexes in Fact and Fiction*
Heather Murray, *Come, bright Improvement! The Literary Societies of Nineteenth-Century Ontario*
Joseph A. Dane, *The Myth of Print Culture: Essays on Evidence, Textuality, and Bibliographical Method*
Christopher J. Knight, *Uncommon Readers: Denis Donoghue, Frank Kermode, George Steiner, and the Tradition of the Common Reader*
Eva Hemmungs Wirtén, *No Trespassing: Authorship, Intellectual Property Rights, and the Boundaries of Globalization*
William A. Johnson, *Bookrolls and Scribes in Oxyrhynchus*
Siân Echard and Stephen Partridge, eds, *The Book Unbound: Editing and Reading Medieval Manuscripts and Texts*
Bronwen Wilson, *The World in Venice: Print, the City, and Early Modern Identity*
Peter Stoicheff and Andrew Taylor, eds, *The Future of the Page*
Jennifer Phegley and Janet Badia, eds, *Reading Women: Literary Figures and Cultural Icons from the Victorian Age to the Present*
Elizabeth Sauer, *"Paper-contestations" and Textual Communities in England, 1640–1675*
Nick Mount, *When Canadian Literature Moved to New York*
Jonathan Earl Carlyon, *Andrés González de Barcia and the Creation of the Colonial Spanish American Library*
Leslie Howsam, *Old Books and New Histories: An Orientation to Studies in Book and Print Culture*
Deborah McGrady, *Controlling Readers: Guillaume de Machaut and His Late Medieval Audience*
David Finkelstein, ed., *Print Culture and the Blackwood Tradition*
Bart Beaty, *Unpopular Culture: Transforming the European Comic Book in the 1990s*

Elizabeth Driver, *Culinary Landmarks: A Bibliography of Canadian Cookbooks, 1825–1949*

Benjamin C. Withers, *The Illustrated Old English Hexateuch, Cotton Ms. Claudius B.iv: The Frontier of Seeing and Reading in Anglo-Saxon England*

Mary Ann Gillies, *The Professional Literary Agent in Britain, 1880–1920*

Willa Z. Silverman, *The New Bibliopolis: French Book-Collectors and the Culture of Print, 1880–1914*

Lisa Surwillo, *The Stages of Property: Copyrighting Theatre in Spain*

Dean Irvine, *Editing Modernity: Women and Little-Magazine Cultures in Canada, 1916–1956*

Janet Friskney, *New Canadian Library: The Ross-McClelland Years, 1952–1978*

Janice Cavell, *Tracing the Connected Narrative: Arctic Exploration in British Print Culture, 1818–1860*

Elspeth Jajdelska, *Silent Reading and the Birth of the Narrator*

Martyn Lyons, *Reading Culture and Writing Practices in Nineteenth-Century France*

Robert A. Davidson, *Jazz Age Barcelona*

Gail Edwards and Judith Saltman, *Picturing Canada: A History of Canadian Children's Illustrated Books and Publishing*

Miranda Remnek, ed., *The Space of the Book: Print Culture in the Russian Social Imagination*

Adam Reed, *Literature and Agency in English Fiction Reading: A Study of the Henry Williamson Society*

Bonnie Mak, *How the Page Matters*

Eli MacLaren, *Dominion and Agency: Copyright and the Structuring of the Canadian Book Trade, 1867–1918*

Ruth Panofsky, *The Literary Legacy of the Macmillan Company of Canada: Making Books and Mapping Culture*

Archie L. Dick, *The Hidden History of South Africa's Book and Reading Cultures*

Darcy Cullen, ed., *Editors, Scholars, and the Social Text*

James J. Connolly, Patrick Collier, Frank Felsenstein, Kenneth R. Hall, and Robert Hall, eds, *Print Culture Histories beyond the Metropolis*

Kristine Kowalchuk, *Preserving on Paper: Seventeenth-Century Englishwomen's Receipt Books*

Ian Hesketh, *Victorian Jesus: J.R. Seeley, Religion, and the Cultural Significance of Anonymity*

www.ingramcontent.com/pod-product-compliance
Lightning Source LLC
LaVergne TN
LVHW040151080826
844660LV00014B/921/J

* 9 7 8 1 4 4 2 6 4 5 7 7 6 *